OXFORD
UNIVERSITY PRESS

ASPIRE
SUCCEED
PROGRESS

exam success

in

GEOGRAPHY

for Cambridge IGCSE® & O Level

Oxford excellence for Cambridge IGCSE® & O Level

OXFORD
UNIVERSITY PRESS

Great Clarendon Street, Oxford, OX2 6DP, United Kingdom

Oxford University Press is a department of the University of Oxford. It furthers the University's objective of excellence in research, scholarship, and education by publishing worldwide. Oxford is a registered trade mark of Oxford University Press in the UK and in certain other countries

First published in 2019

British Library Cataloguing in Publication Data
Data available

978-0-19-842793-3

10

Paper used in the production of this book is a natural, recyclable product made from wood grown in sustainable forests.
The manufacturing process conforms to the environmental regulations of the country of origin.

Printed and bound by CPI Group (UK) Ltd, Croydon CR0 4YY

Acknowledgements

The publisher and authors would like to thank the following for permission to use photographs and other copyright material:

Cover: biletskiy/Shutterstock; **p7:** OUP/Photodisc; **p53:** Robert Gilhooly/Alamy Stock Photo; **p73, 78, 79:** Courtesy of Muriel Fretwell; **p81(t):** Hugh Threlfall/Alamy Stock Photo; **p81(b):** David J. Green - technology/Alamy Stock Photo; **p106:** Shutterstock; **p154:** Corbis RF/Alamy Stock Photo; **p177:** Vadim Sadovski/Shutterstock.

Artwork by Aptara Inc, Barking Dog Art, Q2A Media Services Inc, Steve Evans, and Hart McLeod.

We are grateful to the authors and publishers for use of extracts from their titles and in particular for the following:

Data from https://data.worldbank.org/indicator/. © 2018 The World Bank Group. All Rights Reserved.

Data from www.homeaffairs.gov.au/about/reports-publications/research-statistics/live-in-australia/migration-programme. Australian Government Department of Home Affairs. Licensed under a Creative Commons Attribution 4.0 International licence.

Although we have made every effort to trace and contact all copyright holders before publication this has not been possible in all cases. If notified, the publisher will rectify any errors or omissions at the earliest opportunity.

This Exam Success Guide refers to the Cambridge IGCSE® Geography (0460), Cambridge IGCSE® (9-1) Geography (0976) and Cambridge O Level Geography (2217) Syllabuses published by Cambridge Assessment International Education.

Contents

Introduction 4

Theme 1: Population and settlement

1 Population dynamics (1) 8
2 Population dynamics (2) 12
3 Migration 18
4 Population structure 23
5 Population density and distribution 30
6 Settlements and service provision (1) 35
7 Settlements and service provision (2) 39
8 Urban settlements 43
9 Urbanization 49

Theme 2: The natural environment

10 Earthquakes 53
11 Volcanoes 57
12 Rivers (1) 61
13 Rivers (2) 66
14 Coasts (1) 70
15 Coasts (2) 74
16 Weather (1) 80
17 Weather (2) 86
18 Climate and natural vegetation: Equatorial climate and tropical rainforest 94
19 Climate and natural vegetation: Hot desert 100

Theme 3: Economic development

20 Development 107
21 Food production 113
22 Industry 119
23 Tourism 125
24 Energy 131
25 Water 137
26 Environmental risks of economic development (1) 140
27 Environmental risks of economic development (2) 146

Theme 4: Geographical skills

28 Survey (topographic) maps 154
29 Physical and human features on survey maps 160
30 Photographs, field sketches, data tables and graphs 168

Theme 5: Coursework skills

31 Coursework: Planning investigations 177
32 Coursework: Data collection 184
33 Coursework: Presentation, analysis and evaluation 191

Glossary 200
Revision planners 207

Please go to **www.oxfordsecondary.com/esg-for-caie-igcse** for:

- Answers
- Mark schemes
- Web only exam questions and mark schemes

Introduction

Matched to the latest Cambridge assessment criteria, this in-depth Exam Success Guide brings clarity and focus to exam preparation with detailed and practical guidance on raising attainment in IGCSE® & O Level Geography.

This Exam Success Guide can be used alongside *Complete Geography for Cambridge IGCSE® & O Level* and contains links to the Student Book where necessary.

This Exam Success Guide:

- Is **fully matched** to the latest Cambridge IGCSE® & O Level syllabuses
- Includes comprehensive **recap** and **review features** which focus on key course content
- Equips you to **raise your grade** with sample responses and examiner commentary
- Will help you to **understand exam expectations** and avoid common mistakes with **examiner tips**
- **Apply knowledge** and test understanding via **exam-style questions**, with answers available online
- Is perfect for use alongside the *Complete Geography for Cambridge IGCSE® and O Level Student Book* or as a standalone resource for independent revision

Key revision points are included as follows:

- **Key ideas:** at the start of every section. This summarises key things you need to know for each topic.
- **Recap:** each chapter recaps key content and theory through easy-to-digest chunks and visual stimulus. Key terms appear in bold and are defined in the Glossary.

- **Apply:** targeted revision activities are written specifically for these guides, which will help you to apply your knowledge in the exam paper. These provide a variety of transferrable exam skills and techniques. By using a variety of revision styles you'll be able to cement your revision.

- **Review:** throughout each section, you can review and reflect on the work you have done, and find advice on how to further refresh your knowledge. This will include references back to the Student Book or link synoptically to other sections in your Exam Success Guide.

- **Analysis:** Strengthen exam performance through analysis of sample student answers and examiner responses.
- **Exam tips and Common errors:** include particular emphasis on content and skills where students commonly struggle. The tips give details on how to maximise marks in the exam.

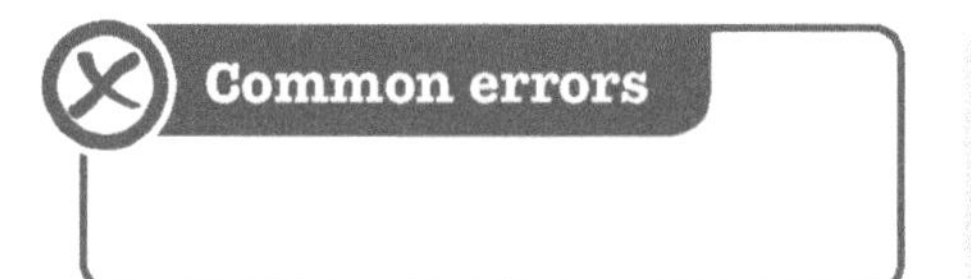

Exam tip

- **Raise your grade:** can be found at the end of each chapter. This section includes answers from candidates who didn't achieve maximum marks and advice on how to improve answers.

You are also encouraged to build a record of essential key terms, and either track your revision progress or use the guidelines to indicate topics you are more or less confident about.

Use the revision planners on pages 207 and 208.

Monday	Tuesday	Wednesday	Thursday	Friday	Saturday	Sunday

Please go to **www.oxfordsecondary.com/esg-for-caie-igcse** for:
- Answers
- Mark schemes
- Web only exam questions and mark schemes

How will you be assessed?

All candidates take three papers for their IGCSE Geography qualification. All candidates take Paper 1 and Paper 2:

Paper 1 Geographical themes (1 hour 45 minutes)

There are a total of 75 marks available, worth 45% of your IGCSE. The paper consists of short-answer levels of response and includes 7-mark level-marked questions.

You must answer one question in each section (A, B, and C) – three questions in total, each worth 25 marks.

Paper 2 Geographical skills (1 hour 30 minutes)

There are a total of 60 marks available, worth 27.5% of the IGCSE. The paper consists of short-answer levels of response questions. You must answer all questions.

And, *either*:

Component 3 Coursework

There are a total of 60 marks available, worth 27.5% of the IGCSE. You must complete an individual investigation based on a question or issue set by your teachers.

You are advised to write up to 2000 words. The investigation will be marked by your teachers and moderated by Cambridge.

Or:

Paper 4 Alternative to coursework (1 hour 30 minutes)

There are a total of 60 marks available, worth 27.5% of the IGCSE. The paper consists of two compulsory questions, consisting of a series of written tasks.

Population and settlement

Chapters 1 to 9 are designed to help you know and understand the rapid increase in the world's population, the link between population and resources, the increase in population migration around the world, the types and location of settlements and the changes taking place in settlements around the world today.

There will be two questions on *Population and settlement* in the *Cambridge IGCSE® or O Level* Paper 1, but there will also be at least one question in Paper 2. You may also find questions on settlement in Paper 4. Although population and settlement are separate topics in the syllabus, they are strongly linked. For example, population growth and migration have led to urbanization and have affected settlements both in rural and in urban areas.

1 Population dynamics (1)

Key ideas

- The total population of the world has risen continuously to reach 7.5 billion in 2017.
- World population totals are expected to peak at about 10 or 11 billion in the late 2080s.
- The world's rate of population change accelerated to reach its highest rate (2.2%) in the 1960s.
- Between the 1960s and 2011 the rate of world population growth halved and continues to reduce.
- At the same time, the total world population continued to rise but at a slower rate.
- The world population is increasing because overall birth rates are higher than death rates.

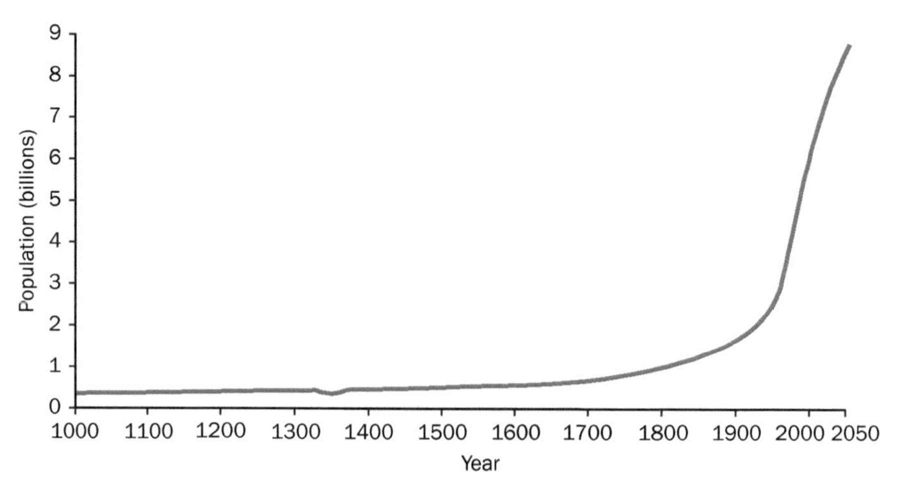

▲ **Fig. 1.1** *Can you spot the results of the Black Death plague and the influences of better nutrition, healthcare and clean water supplies?*

Reasons for the rapid increase in the world's population

Many less economically developed countries (LEDCs) in central Africa, such as Niger, and some in south-east Asia, such as Cambodia, have rapidly increasing populations because they are in the earlier stages of economic development. **Birth rates** either remain high or are falling slowly, but **death rates** are falling rapidly and have been doing so over a longer period. During the period of the world's most rapid increase, many countries had rapid population increases and their increases were more rapid than before.

High birth rates caused by:	Reducing death rates caused by:
Culture In many cultures: • women have large numbers of children (high fertility rates); this is especially so in polygamous societies • the inferior status of women results in them having little say in decisions about family size • the more children a man has, the more prestigious he is considered • the desire for a son is of great importance, so more children will be born until a son is produced • children are needed to support parents in their old age • early marriage is traditional, so there are more childbearing years.	**Healthcare** There is better health resulting from: • improvements in medicines, vaccinations, medical training and doctor/nurse-patient ratios • more clinics and hospitals • the eradication of some illnesses that used to cause epidemics • better nutrition and greater awareness of healthy diets and lifestyles • better access to food supplies, partly resulting from better farming techniques and new higher yielding varieties of cereals • increased provision of clean water supplies and sanitation, especially in rural areas.

Common errors

- Be clear about the difference between *total* population growth and *rates* of population growth.
- Describing population change when the question asks for reasons for it and vice versa is a waste of valuable examination time. Always obey the key command word(s) in a question.

Common errors

Remember that a fall in the growth rate will not result in a fall in the total population until after the growth rate becomes negative.

High birth rates caused by:	Reducing death rates caused by:
Economic reasons • Children work on family farms or to increase the family income. • Poverty prevents people being able to purchase or access contraception. • Countries with a poor economy are unable to provide enough birth control clinics. • Poverty leads to malnutrition and higher infant and child mortality rates, so people have more children to ensure that some survive.	**Economic reasons** As a country develops economically: • more birth control facilities can be made accessible to more people. • internet coverage increases, enabling greater awareness of the benefits of smaller families and of differences between societies.
Religion Catholic and Muslim families usually have more children as the use of contraceptives is discouraged and natural contraception favoured.	
Demographic reasons Countries with a high proportion of females of childbearing age have higher birth rates.	**Demographic reasons** Countries with a high proportion of elderly have higher death rates.

Exam tip

Make sure you know case studies of a country with a high **rate of natural population growth**, such as Niger, and one with a low **rate of natural population growth** or population decline, such as Japan.

Apply

Draw three graphs to show the patterns of birth rates and death rates which result in:

(a) a rapidly increasing population

(b) a low rate of population growth

(c) a slowly declining population.

Use these as headings for each diagram.

Below each graph (a), (b) and (c), complete a table to include: name of an example country, its economic status (e.g. LEDC) and influences on its birth and death rates.

See www.oxfordsecondary.com/esg-for-caie-igcse for the 'Apply' task answers.

Review

- The rate of increase in world population has slowed since the 1960s.
- Differences in social, economic and **demographic factors** cause contrasts in **natural population** change between countries.
- Where women have a low status in their cultures, birth rates tend to be high.
- Death rates vary with differences in economic development and demography.

Population policies of governments also influence birth rates, as described in Chapter 4.

There are case studies of Niger, a country with a high **rate of natural population growth**, on pages 11–12 of *Complete Geography for Cambridge IGCSE and O Level*. Also, on page 16, there is a case study of Japan, a country with a negative growth rate.

Recap

Different parts of the world have contrasting rates of population change.

- LEDCs have high rates of natural increase because birth rates are considerably higher than the death rates, which are falling.
- At first, MEDCs have low rates of natural increase because their birth rates are only slightly higher than their death rates, and both are low.
- MEDCs that are in a late stage of development have declining populations because they have a high proportion of elderly, so death rates are higher than birth rates.

Sample question

1. Study the table giving information for 2017 about the population of Estonia, a small European country.

	/000	%
Birth rate	10.1	1.01
Death rate	12.6	1.26

(a) What is meant by *birth rate?* **[1 mark]**

(b) Calculate the natural population growth rate for Estonia. Show your working. **[2 marks]**

(c) Estonia has had a similar natural population change since 1992. State **two** problems for Estonia which result from this change and give **one** way in which the country could reduce the problems. **[3 marks]**

Analysis

✓ You need to be able to define terms that are in the syllabus, without using words that are included in the term being defined.

✓ You should be able to make simple calculations accurately or with a calculator.

✓ You must put a clear minus sign if the answer is a negative.

✓ This question gives an example of a country in which the natural population has been decreasing, but it will not necessarily decline in future years.

✓ In (c) there are two tasks to answer, not just one, as this question tests your knowledge of problems of, and a possible solution to, negative natural population 'growth' for a country.

✓ Natural population change is often used instead of natural population growth. The calculation is the same whether or not the answer is a positive growth or a decline.

✓ The command word 'state' means that you should give a clear answer. It does not expect further detail about the answers.

Student answer

(a) The number of births per 1000 in a population. 0 marks

(b) 10.1 − 12.6 = −2.5. 1 mark

(c) Estonia should increase the birth rate and try to attract immigrants. 0 marks

Total: 1 out of 7

See www.oxfordsecondary.com/esg-for-caie-igcse for the mark scheme for this question.

Examiner feedback

(a) As the definition is of a *rate*, it must refer to the time over which the number of births is measured, so cannot gain a mark unless 'per year' is stated.

(b) The calculation is correct, but the units are missing.

(c) The question asked for *one* solution and two are given, so only the first is acceptable. However, both are too vague for a mark. Give detail, such as how the method will be encouraged or how it should alleviate the difficulty.

2. Study the diagram showing typical population changes over time in many countries.

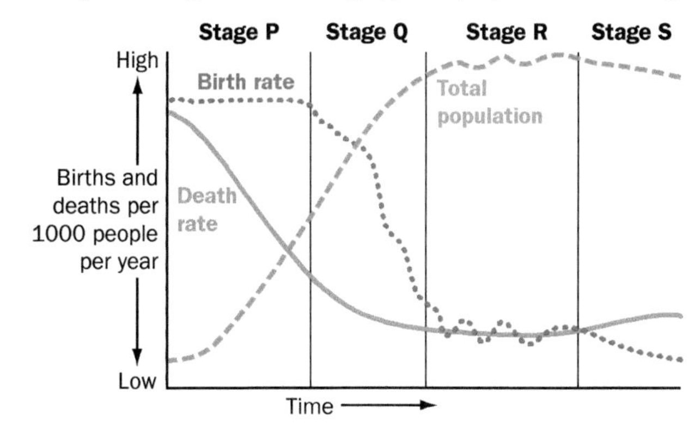

(a) Compare the changes in a country's total population during Stages P and S. **[1 mark]**

(b) Describe and explain the change in the total population of a country during Stage R. **[2 marks]**

(c) Describe the problems for people and governments that are caused by the total population change in Stage P. **[4 marks]**

Analysis

✓ This question tests your knowledge of how changing birth and death rates result in differences in natural population change between countries.

✓ It considers the consequences of one type of population change.

✓ It examines your ability to make a comparison.

See www.oxfordsecondary.com/esg-for-caie-igcse for the mark scheme for this question.

Student answer

(a) Population is lower in Stage P than Stage S. [0 marks]

(b) Population is stable in Stage R because the birth rate has fallen to the same level as the death rate and they remain like this through nearly all the stage. [1 mark]

(c) There are not enough jobs for all the people, and many households find it difficult to support their large number of children. They may over-cultivate and impoverish the soils on their subsistence plots, which will cause less yield in future years. Hunger and eventually starvation can result. There is little hope of government help. [3 marks]

Total: 4 out of 7

Examiner feedback

Key words in the question were missed in both parts (a) and (b), resulting in low marks. In (a), the question asked for a comparison, which this answer gives, but it asked for a comparison of the *changes*, which this answer does not give. Again, in part (b) the question asked for a description of a change and this has not been stated, but a mark is gained for the reason for the change from a rising population to a stable population. In part (c) there is a reference to the government, but it is too vague. It should be explained that the finances of the government itself are too low to help the people, because few workers produce little taxation income for it. This low total mark could be avoided by reading the question more carefully, underlining the key words 'describe', 'explain' and 'change(s)' in it and doing what they say.

2 Population dynamics (2)

Key ideas

- **Over-population and under-population describe the relationship between population and resources in a country or area.**
- **They describe a lack of balance between the needs of the population for resources and the resources it has.**
- **The ideal balance is when the entire population of a country has a good standard of living by using the country's resources to build a strong economy.**

Common error

Remember that a country with a large population is not necessarily **over-populated** and one with a low population is not always **under-populated**. The balance with resources must be considered.

Over-population and under-population

Causes of over-population

- Population pressure on **resources** could be the result of a very favourable physical environment which has attracted settlers.
- The resources attracted more settlers than they could support.
- A large influx of immigrants can put a strain on resources.

Causes of under-population

- Lack of population to fully use the resources may be because of a hostile physical environment.
- The environment might be hostile to people, but rich in geology, such as the tundra of northern Canada, which has abundant minerals and rocks.
- A country may become under-populated if many inhabitants emigrate, leaving insufficient labour to fully develop its resources.

Characteristic	Over-populated country e.g. Bangladesh	Under-populated country e.g. Australia
Unemployment	High	Low
Energy and mineral resources	Few	Many
GDP (gross domestic product – a measure of wealth)	Small	Large
Export earnings	Low	High
Standard of living	Low	High
Net population movement	Out of the country	Into the country
Main sector of employment	Primary (agriculture)	Tertiary (services)
Main problems caused by the imbalance between population and resources	Shortage of food leading to deforestation, soil degradation, poverty, overcrowded urban areas (e.g. Dhaka), squatter settlements, water, noise and air pollution, crime.	Shortage of workers to develop the resources. Many immigrant workers lead to social conflict in cities (e.g. Brisbane). High cost of importing goods not produced in the country.

Exam tip

Make sure you know a case study of a country that is **over-populated** and also a country that is **under-populated**. Be able to quote some statistics, e.g. the employment and migration rates and the types of resources or lack of them to support your classification of the country.

Key ideas

- HIV/AIDS reduces birth rates and increases death rates.
- The overall effect of these changes is a change in the population size, as it reduces the natural increase of the population to below what it would have been if the virus had not affected it.
- HIV/AIDS is still impacting populations but to a reduced extent, as drugs have now been developed to control it.
- Antiretroviral drugs are often in short supply in LEDCs, partly because they are expensive.

The impact of HIV/AIDS on birth and death rates

How HIV/AIDS impacted the birth rate

- It spread in body fluids, particularly during unprotected sex, causing the deaths of many women of childbearing age.
- Mothers passed the virus on to their babies, so fewer of them lived to childbearing age.
- HIV spread rapidly in LEDCs in southern Africa, such as Botswana, Swaziland and Lesotho, especially where poverty encouraged prostitution and contraception was not available. It spread so rapidly that only 13 years after it was first identified in Botswana one in four of Botswana's population between the ages of 15 and 49 was HIV positive.

How HIV/AIDS impacted the death rate before the development of antiretroviral drugs

- Botswana's death rate was seven times higher than it would have been without AIDS.
- Its infant mortality rate was three times higher than before AIDS.
- Life expectancy fell from 72 years to only 34 years.

Botswana's natural increase of population fell from 3.3% to 1.1% in the ten years to 2011. However, the effects of the antiretroviral drugs are now leading to a slow increase in the birth rate and a decrease in the death rate. There is a consequent increase in natural population growth, but full recovery will take a long time.

Recap

- An over-populated area has too many people to be supported to a good standard of living by its resources.
- An under-populated area has too few people to fully use its resources to achieve a good standard of living.
- HIV/AIDS reduces natural population growth for many years by lowering the birth rate and increasing the death rate.
- Antiretroviral drugs have had a limited success in southern Africa because of the high cost.

Apply

1. List four characteristics of a country that are strong indicators of it being over or under-populated. Present your answer in a table and name an example of each country.
2. Make another table summarizing two consequences of over-population for a rural area compared with a town.
3. Describe the ways in which HIV/AIDS has affected the birth rate, death rate and natural population growth of a named country. Include one supporting statistic for each of these population characteristics.

See www.oxfordsecondary.com/esg-for-caie-igcse for the answers to the 'Apply' task.

Review

- Over-population can be caused by the depletion of resources and a large influx of migrants, under-population by the discovery of new resources and emigration, provided the population remains the same.
- Look at the graph for Botswana in the answer to the 'Apply' task to help you remember the effects of HIV/AIDS on natural population change. Pages 6–7 of *Complete Geography for Cambridge IGCSE and O Level* have further detail about HIV/AIDS in Botswana.

There are also case studies of Bangladesh (an over-populated country) and Australia (an under-populated country) on pages 9–10 of *Complete Geography for Cambridge IGCSE and O Level.*

Raise your grade

Sample question

1. Study the graph of birth and death rates in Botswana, an LEDC in southern Africa.

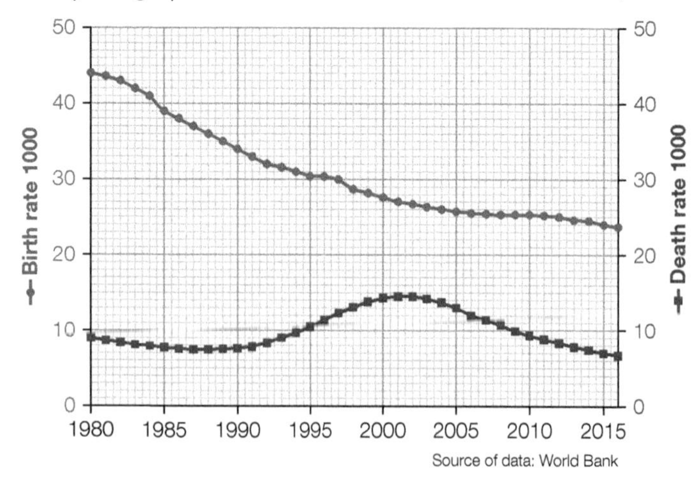

(a) The HIV/AIDS epidemic reduces birth rates. Explain why this effect is difficult to see on a graph of an LEDC country, such as Botswana. **[1 mark]**

(b) Describe the impact of HIV/AIDS on the death rate of Botswana, as shown on the graph. **[2 marks]**

(c) Women of childbearing age are most affected by the virus in some societies. Explain why this is so and why HIV/AIDS reduces the birth rate for many years. **[3 marks]**

Analysis

✓ This question is a reminder that there are several influences on birth and death rates, of which HIV/AIDS is one.

✓ The answer to the second question is derived from information on the graph, but the main requirement for the first question is knowledge of birth rate trends in LEDCs.

✓ An impact on the size of the population in one generation has a knock-on impact on the demographics of the next generation.

✓ The impact of HIV/AIDS on the death rate is similar to that of any other epidemic.

✓ The question highlights how inequality leads to the particular vulnerability of many women in some societies.

✓ In order to demonstrate how something has had an impact on a graph it is necessary to state a change.

✓ Two of these questions ask for explanations. That means that you must say why.

✓ It is important to note that the last question has two requirements to be answered before full marks can be obtained.

Student answer

(a) The birth rate was declining before AIDS was diagnosed, so it would only be possible to notice a decline caused by HIV/AIDS if there was a sudden larger rate of decline, which is not the case. [1 mark]

(b) The death rate line goes up after having gone down since the start of the period. [0 marks]

(c) Women are subservient to men in some societies and will be unable to demand use of contraception. Early marriage is traditional in many societies, so girls are exposed to the virus from early adulthood. In societies where traditional healers are used, they cannot survive without medical treatment. [2 marks]

Total: 3 out of 6

Examiner feedback

Part (a) is an excellent detailed explanation. In part (b) the question is about the death rate, but the answer given is about a line on a graph. Always translate lines and symbols on diagrams into reality.

Two good points are made in part (c) about how women of childbearing age in some societies are made more vulnerable to HIV/AIDS than any other section of the society. However, the last demand of the question has not been answered.

See www.oxfordsecondary.com/esg-for-caie-igcse for the mark scheme for this question.

Sample question

2. Study the information for 2016 about Kazakhstan, a country in Central Asia. GDP (Gross Domestic Product) is a measure of wealth.

Kazakhstan	
Area (km)	2717000
Population	18.3 million
Labour force	9 million
Unemployment	4.9%
Migration rate	+0.4 /000 people
Resources	Many, including natural gas, oil, uranium, copper and zinc. Some grain and livestock products
Main employment sector	62% work in services
Economy	
GDP (US$ per person)	$26000 (moderate)
Exports	Oil and oil products, natural gas, ferrous metals, chemicals and machinery
Value of exports (US$ per year)	$35 billion
Imports	Machinery, metal products and foodstuffs
Life expectancy	70

(a) Identify **one** characteristic in the table that suggests that Kazakhstan may not be an under-populated country. **[1 mark]**

(b) Identify **two** characteristics that suggest that Kazakhstan is an under-populated country. **[2 marks]**

(c) Describe the problems for a country if it is unable to fully exploit its resources. **[4 marks]**

Analysis

✓ This question tests the concept of under-population which relies on a country having too low a population to fully exploit its resources to support its population to the best possible living standard.

✓ In reality, countries are so varied that they are unlikely to fulfil all the characteristics of being under- or over-populated, so it is necessary to select answers with care from the information supplied.

✓ As a country develops, it can change from being under-populated to being over-populated.

✓ The command word 'identify' means that you should name or select what is correct.

Student answer

(a) Under-populated countries usually have much higher immigration of migrant workers into the country than the net migration rate of 0.4%, which is extremely small. However, there may be political restrictions on immigration or other factors deterring immigrants. [1 mark]

(b) Kazakhstan has a small population in a large area. It exports many resources and imports products, such as metal products made from resources it exports, so it must lack the necessary labour to manufacture them (as it has the energy needed). A low unemployment rate of 4.9% suggests it needs to attract migrant workers to exploit its resources more fully. [2 marks]

(c) If a country cannot fully exploit its resources, it will not earn the maximum amount of income for the economy. That will limit what it can spend and its people will not have as good a living standard as would have been possible in a richer country. [2 marks]

Total: 5 out of 7

See www.oxfordsecondary.com/esg-for-caie-igcse for the mark scheme for this question.

Examiner feedback

Part (a) is a good answer, although the second sentence is irrelevant and would waste valuable time in an examination.

In part (b), a small population in a large area is not an indication of under-population unless it is related to the resources available. Better evidence is the small labour force and large number of valuable resources. Low unemployment is valid, as it suggests a shortage of suitable workers to turn raw materials into more valuable finished products.

In part (c), two points are well made, but the reference to being limited in what the country can spend is too vague. It is better to be specific by suggesting some of the ways in which governments improve the standard of living of their populations.

3 Migration

Key ideas

- **Migration affects the total population growth (or decline) of an area.**
- **Migration has positive and negative effects on the areas of origin and destination areas of the migrants.**
- **Migration has positive and negative effects on the migrants themselves.**
- **The majority of migrants are young male adults.**

Common error

Remember that natural population change does not include the effect of migration.

The influence of migration on the total population of an area

Total population = natural population + **net migration**

The full calculation:

(birth rate – death rate) + (number of **immigrants** – the number of **emigrants**)

Types of migration

Net migration of an area can be **positive** (more people come in than leave) or **negative** (more leave than move in). **Migration** can be:

- internal (within a country) or international (between countries)
- voluntary by **economic migrants**, or involuntary by **refugees**
- temporary, e.g. seasonal migrants (who move for one season to work), or permanent.

Internal migration can be:

- rural to urban migration which is more common in LEDCs
- urban to rural migration which is more common in MEDCs
- urban to urban, such as when people get a job in another town
- transmigration (an involuntary movement of people by the government of Indonesia from over-populated islands to relatively under-populated islands).

International migrations are mainly:

- to neighbouring countries *or*
- from LEDCs to MEDCs.

Causes of migration

Push factors attract people to an area and pull factors cause people to leave an area.

Push factors	Pull factors
Unemployment	Employment opportunities
Low wages	Higher wages than at home
Poor healthcare	Better healthcare
Poor educational opportunities	Better educational opportunities
Low standards of living	Better standards of living
Few entertainment or leisure facilities	More entertainment and leisure facilities
War, crime and persecution	Peace, less crime and a tolerant society
Natural disasters	Safer areas with provisions for refugees
Homesickness	To be near friends and family

Common errors

Remember that many pull factors are the direct opposites of push factors, so do not use more than one of each pair in explanation of why migration occurred.

The impacts of migration

Impacts of emigration on the area of origin

- As most economic migrants and many political refugees are young able-bodied men, their home communities lose workers.
- Families lose fathers, who are often providers.
- Women are over-burdened.
- Fewer workers cause economic decline.
- The most educated usually migrate in voluntary migration, so health and other services decline because of shortages of skilled teachers, doctors and nurses.

Impacts of immigration on the destination area

- There is an immediate need to provide food, clothing and shelter.
- There is an increase in total population.
- The increase continues for generations if the immigrants traditionally have higher fertility rates.
- Rapid population growth puts a strain on housing, education and health services, especially maternity services.
- Immigrants tend to live together, instead of integrating into the society to which they moved.
- Culture clashes and racial tension between the immigrants and local people can result from the lack of normal social interaction. There can be a lack of understanding by the immigrants of the culture and laws of the host country. The local people might also be hostile towards the immigrants because of a lack of understanding of their culture.
- Immigrants are willing to work for lower wages, depressing incomes and causing unemployment and resentment among the local workers about increased competition for certain types of jobs.
- Positive impacts on the receiving countries include a supply of labour to help the economy, especially because immigrants are willing to undertake jobs the locals might not.
- Immigrants who find work pay taxes.
- They create a larger market for local businesses.
- They provide new services, including a diversity of food in ethnic restaurants.
- Immigrants bring new cultural experiences in art, music and literature, as well as creating a better understanding of different cultures.

Exam tip

Make sure you know a case study of international migration and are able to supply some specific details about it.

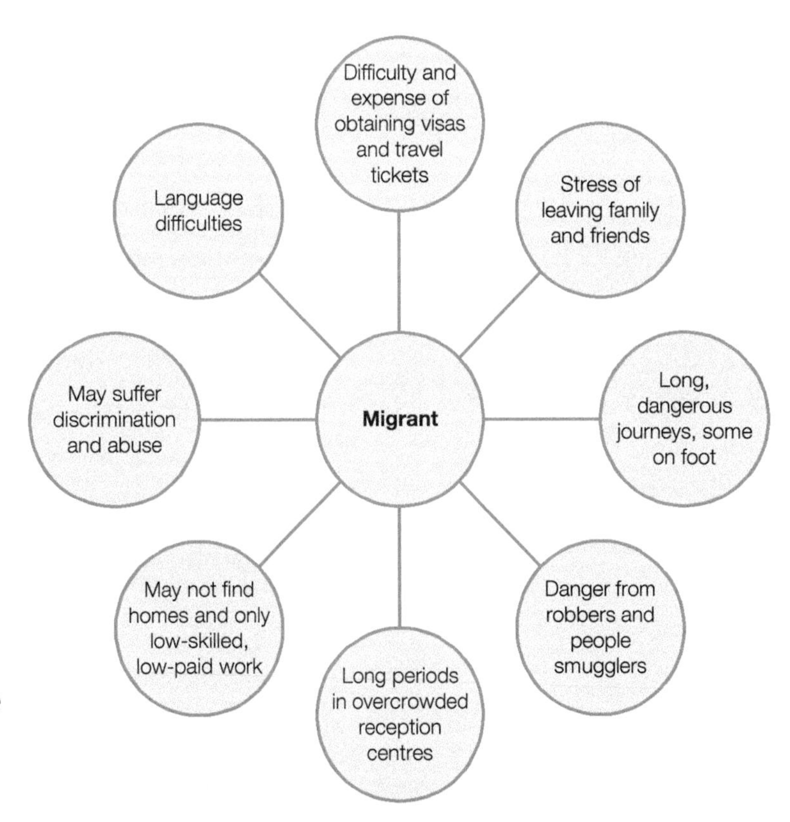

▲ **Fig. 3.1** *Impacts on migrants*

Recap

- A positive net migration adds to the natural population to increase the total population, whereas a negative net migration decreases it.
- Push factors are reasons for emigration and pull factors attract immigration. Push factors include war, discrimination, famine and poverty.
- Migration has both positive and negative effects for the migrants and for the countries of origin and destination.

Apply

This task is complemented by Question 1 in Chapter 4.

> In 2014 Bahrain, a Middle Eastern country, encouraged immigration to improve its infrastructure and development. At the height of the building boom it had 730 000 immigrants and a national population of only 600 000. There were as many Asian economic immigrants as Bahraini nationals, 70% of them from India, Bangladesh and Pakistan. Female nurses came mainly from the Philippines.

Comment on the migration to Bahrain under the following headings:

- Effects of economic migration on the total population of Bahrain.
- Reasons for migration from their origin (home countries).
- The impact of this emigration on their home countries.
- The problems large numbers of immigrants could cause for Bahrain.
- Problems of migrating for the migrants themselves.

See **www.oxfordsecondary.com/esg-for-caie-igcse** for the answers to the 'Apply' task.

Review

- There are now many economic migrants and refugees. Education and the media give people in LEDCs a glimpse of a better life in MEDCs, where rapidly ageing populations need more workers. Cities attract large numbers of job-seeking migrants, so some city dwellers move out, putting pressure on the surrounding countryside.
- **Mass immigration** raises the birth rate of MEDCs, putting pressure on services, because the culture of many immigrant groups is to have large families.

There are case studies of migrations on pages 19–23 of *Complete Geography for Cambridge IGCSE and O Level* and Chapter 2 details some of its consequences.

Sample question

1. Study the diagram showing annual population change in the United Kingdom from 1992 to 2016.

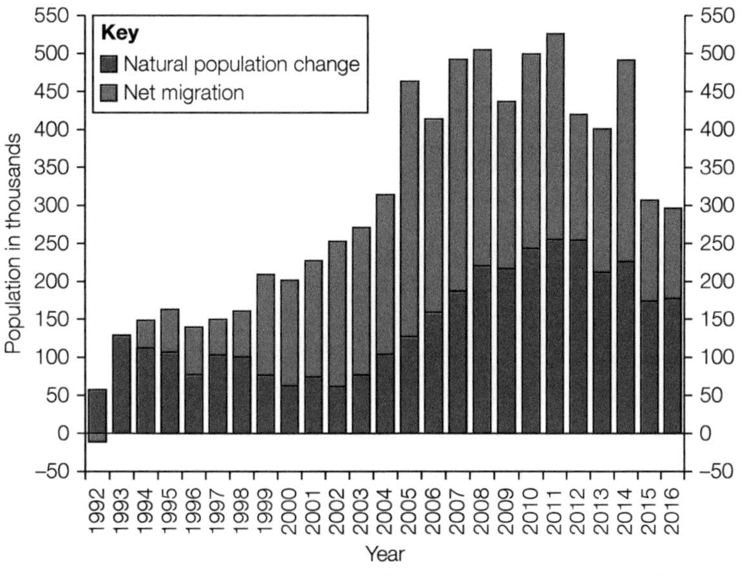

(a) Define *natural population change.* **[1 mark]**

(b) Describe the annual natural population change in the UK between 1992 and 2011. **[2 marks]**

(c) Use the graph to suggest why immigration was encouraged by the UK government in the first half of the period shown. **[4 marks]**

Analysis

✓ The diagram for this question is a compound bar graph, on which the parts that make up the total are shown one on top of the other. It is easy to read values for the lower bars. The values of the upper bars can be obtained in two ways. Either take the lower value from the value of the total of the bar; or measure the length of the top bar on the edge of a piece of paper, put it along the vertical scale line, with one end at zero, and read off the value of the top of the bar. The second method is often easier and less prone to error than calculating.

✓ This graph has a small negative net migration value at the start of the time period which is easy to miss unless the graph is studied carefully.

✓ The graph illustrates the difference between natural population change shown by the lower bars and total population change, which includes net migration.

✓ The graph demonstrates how migrant populations can rapidly influence the birth rates, as over the period shown more babies were born to migrants than to non-migrants.

Student answer

(a) Natural population change is birth rate minus death rate. [0 marks]

(b) It rose, then stayed the same for two years, then fell a little, then rose and stayed the same for two years, then fell. Then it rose from 2002 at a steady rate to its highest, about 255 000 in 2011. [1 mark]

(c) The UK government had a declining population. This had been the case for at least ten years and its workforce was similarly decreasing. In order to keep its industries and agriculture working well it became essential to attract immigrants into the country. As fewer babies were being born, there would be fewer women of childbearing age in the next generation, so the problem would become greater over time. [3 marks]

Total: 4 out of 7

See www.oxfordsecondary.com/esg-for-caie-igcse for the mark scheme for this question.

Examiner feedback

In part (a) the definition given is more an answer to a question asking how natural population change is calculated than an answer to the question asked. It also lacks reference to a time span.

The first sentence of part (b) is an example of how this type of description of change should *not* be done. A better way would be, 'After a rise from 1992–93, the natural population growth fell overall to its lowest of about 60 000 in 2000, with a minor fluctuation in 1997–98.' The description of the final increase was good, as it used figures from both axes. It is important when analysing changing trends over time shown in graphs to summarize the most important trends, not give a year-by-year account of changes. The overall increase is also worth mentioning, as it ended much higher than it was before 2007.

Part (c) is a good answer, just missing the first point in the mark scheme, which was expected because the question instruction was to 'use the graph'. The graph was used, but this point about negative net migration in one year was another use of it.

Key ideas

- Countries have different types of population structure according to their level of economic development.
- Age-sex pyramids are used to illustrate different types of population structure.

Age-sex (population) pyramids

It is useful to divide an age-sex pyramid into three age groups.

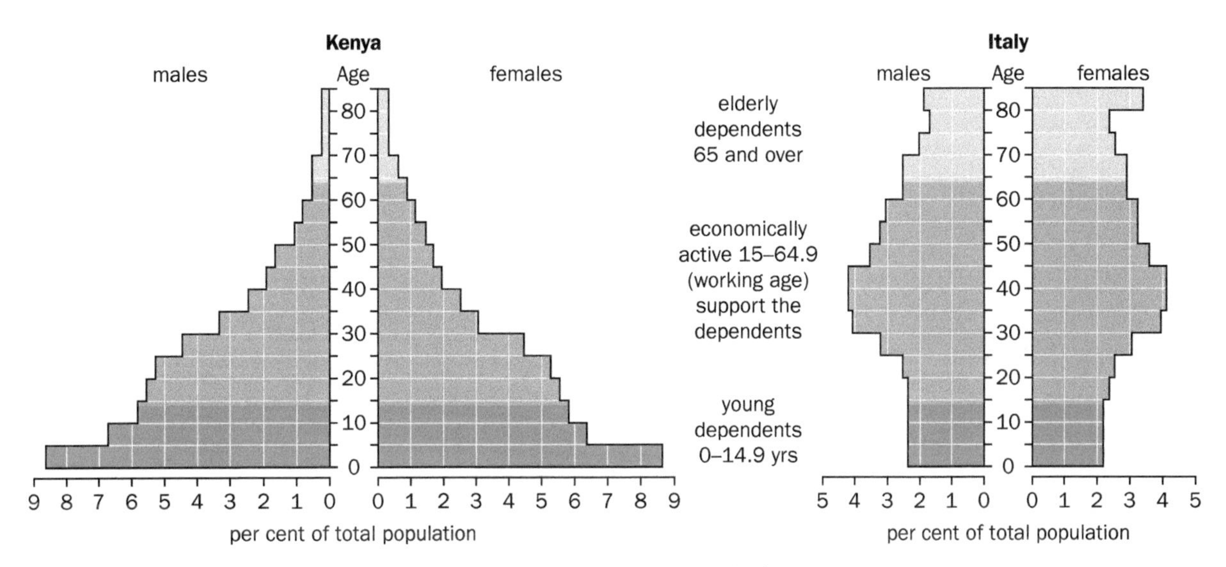

▲ **Fig. 4.1** *Age-sex pyramids. Which one represents an LEDC and which an MEDC?*

Countries with different economic development have different age-sex pyramid shapes.

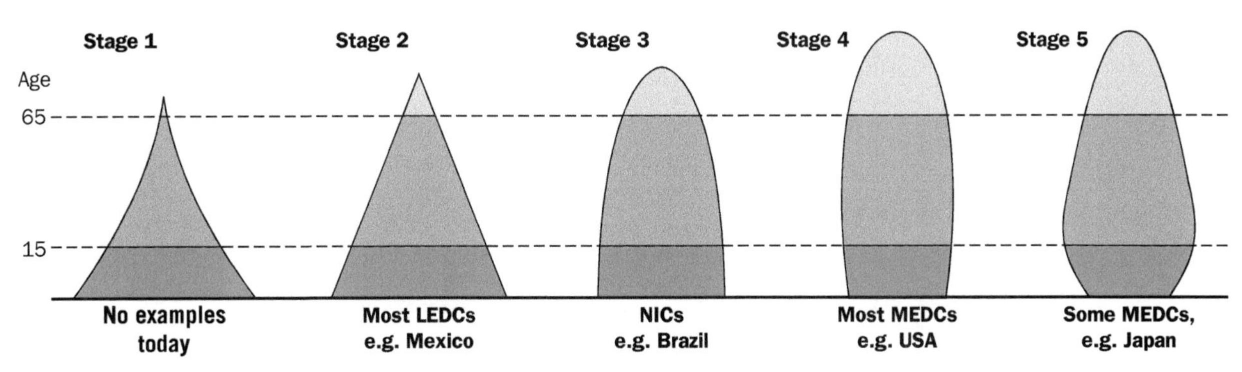

▲ **Fig. 4.2** *As a country develops economically, its population structure is expected to change as shown in these simplified age-sex pyramids. Notice at least four characteristics of the population that change if a country progresses through all the stages.*

Common errors

- If the question asks for a description of population structure, *interpret* the shape of the pyramid to obtain the answer. A description of the shape alone does not describe the structure of the population.
- Always look carefully at the units used on age–sex pyramids. Are the population bars in thousands or percentages? If two pyramids need to be compared, check to see if they are on the same scale and, if not, take care describing which is the larger and smaller.

Pyramid shape	What it indicates
Wide base	Many in the 0–5 age group, suggesting a high birth rate
Narrow base	Few in the 0–5 age group, suggesting a low birth rate
Short	The population has a low life expectancy
Tall	The population has a high life expectancy
Triangular	There are many children and the population numbers fall rapidly with increased age. The population will grow rapidly
Straight-sided	Approximately equal numbers in the age groups
Curves in at the base	Fewest in the youngest age group, indicating a declining birth rate. The population will reduce over time
Pointed top	A high death rate in the elderly population
Dome-shaped top	A low death rate in the larger elderly population (until the age of life expectancy is reached)
Wider on one side than at the same age on the other side	An imbalance between males and females in the age group

Problems associated with different population structures

LEDCs (Stage 2) Low level of economic development	MEDCs (Stages 4 and 5) High level of economic development
1. Increasing numbers of young dependents who need more funding for more healthcare, especially maternity services, schools and food supplies.	**1.** Increasing numbers of elderly dependents who need more funding for more residential homes, social support and medical care, and to pay an increasing number of pensions for many years.
2. Rapid population growth as the increasing number of young reach childbearing age.	**2.** Population growth slows until it eventually declines (in Stage 5).
3. Unemployment as the large numbers born reach adulthood.	**3.** Shortage of workforce as the population ages.
4. Insufficient financial resources to provide for the needs.	**4.** Declining financial resources as less tax is paid by an increasingly smaller economically active population.

Population growth puts a great strain on resources, such as water, food and energy supplies, as well as leading to the loss of natural vegetation and farmland for housing developments. Without care, desertification can occur. Countries with a high **dependency ratio** have many economic and social problems.

Financing the needs of an increasing **dependent population** results in lower living standards for the **economically active** in any country and reduces the funds available for developing the country.

Evaluation of population policies

Policies to reduce the birth rate, population growth and size

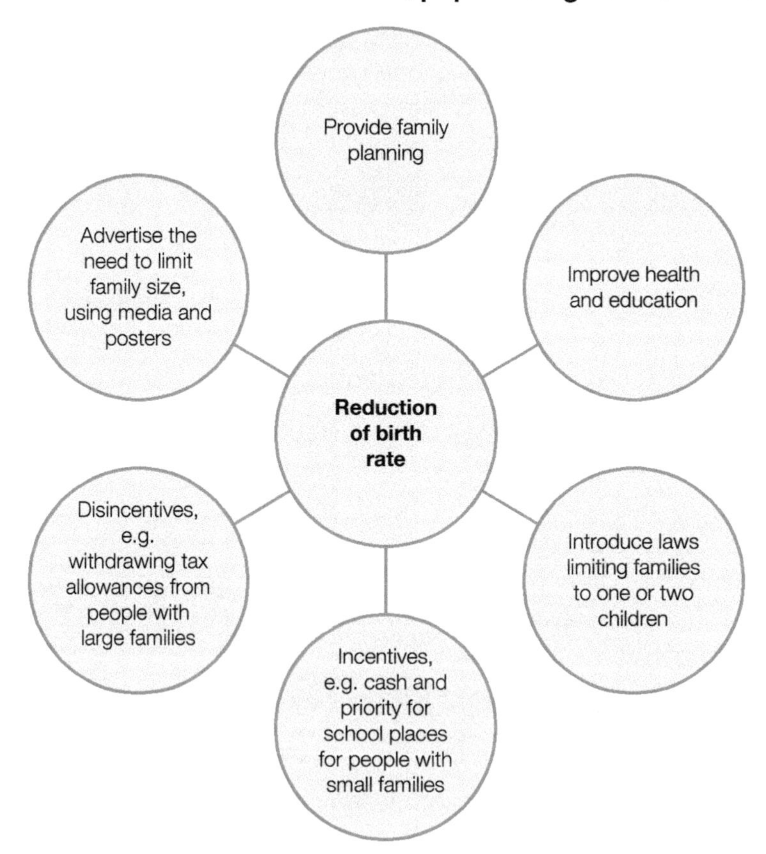

▲ **Fig. 4.3** *Strategies to reduce the birth rate*

China's one-child policy was very successful in reducing the birth rate, but caused many long-term social problems.

Asia and Africa

Providing family planning in poor countries in Africa and Asia has had limited success because:

- there is lack of access to, and awareness of, free contraception in rural areas
- governments have many other problems to spend money on
- in some Muslim and Roman Catholic countries there may be a reluctance to spend money on family planning as having many children is traditionally considered to be desirable
- introducing family planning is difficult in societies where males make all the decisions.

Kerala

Improved health and education has been very successful in Kerala, a state in India, but:

- countries with limited spending power may prioritize other needs, such as combatting HIV/AIDS and malaria
- many countries have shortages of trained doctors and nurses. Education is vital to improving healthcare, but many children receive little schooling.

Policies to increase the birth rate, population size and workforce

Some policies aim to enable mothers to remain in the workforce part-time to earn enough income to maintain a good standard of living, while also bringing up children. They also encourage men to spend more time at home and share the childcare. Incentives used in Sweden include:

- paying a generous benefit for each child born
- giving fathers 13 months' paid leave after the birth of a child
- giving a generous number of paid days off work a year to care for sick children
- providing all-day childcare or all-day schools.

These policies did not have great success until immigrants from the Middle East made Sweden their home after 2012.

France

France gives incentives to encourage couples to have three children, with greater success, as its fertility rate is now one of the highest in Europe. These incentives are:

- the more children a person has, the less tax they pay
- women are paid to stay off work for their third baby's first year
- paid leave for one parent for 36 months after having a child
- subsidized childcare for children under the age of three
- free schooling for children aged over three.

Other policies

More immediate and effective ways of increasing the workforce, which are not always popular, are:

- encouraging immigration
- using robots and technology to replace workers where possible
- raising the retirement age
- giving grants to employers who employ people over pensionable age.

Exam tip

Make sure that you can give details of a country with a high dependent population and are able to supply specific detail, such as the percentage of the population that is dependent.

Recap

Population structure refers to the proportions of the population that are aged 0–14.9 years, 15–64.9, 65 and over, together with the males compared with the females. It indicates the dependency ratio by comparing the non-working population groups with the working population who have to support them. Age-sex pyramids show this information well.

Population structures change as the country develops economically:

- LEDCs have high young dependent populations and few elderly
- MEDCs have few young people, and high elderly populations, which increase the longer the country has been an MEDC.

Governments often adopt policies to try to reduce dependency problems.

Apply

Prepare for a question on the case study of a country with a high dependent population by compiling a list of essential information. You need to be able to quote at least three relevant facts, such as: the dependency ratio for the young or old dependents or overall dependency ratio; the reason for the high dependency e.g. the birth rate and life expectancy; and any policy used to try to reduce the problem, with a brief appraisal of its success or failure. You may be required to state some of the social and economic problems caused by the high dependency and consider the future problems if the population imbalance is not reduced.

See **www.oxfordsecondary.com/esg-for-caie-igcse** for the answer to the 'Apply' task.

Review

Higher dependent populations are a strain on the economy:

- a higher young dependent population needs more maternity services, clinics, schools and food
- a large young generation have even more babies as they reach child-bearing years
- when they become adults there are insufficient jobs for many of them, so poverty spreads
- populations with higher proportions of elderly dependents require more residential homes, hospitals and social care services and receive more pensions from the state.

There are case studies of Niger, a country with a high number of young dependents, on pages 11–12 and 29 of *Complete Geography for Cambridge IGCSE and O Level*. Also, on pages 26–28 there is a case study of Japan, a country with a high number of elderly dependents.

Sample question

1. Study the population pyramid for Bahrain.

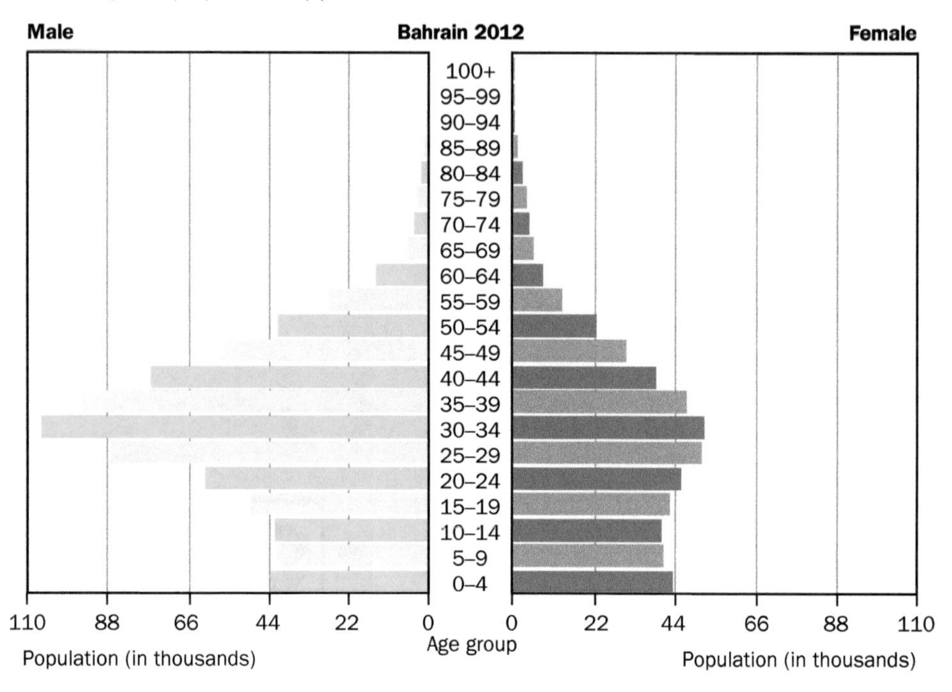

(a) Estimate how many children (0–14 years old) were living in Bahrain in 2012. Choose the answer nearest to yours.

42 000 84 000 120 000 168 000 240 000 **[1 mark]**

(b) Describe **two** ways in which the adult population structure of Bahrain is unusual. **[2 marks]**

(c) Suggest reasons for the main characteristics of the population structure of Bahrain. **[4 marks]**

Analysis

✓ This question tests your knowledge of the meaning of population structure.

✓ It shows an age–sex pyramid shape, which is not typical of one from an LEDC or an MEDC with which you should be familiar. The syllabus states that you may be given resources about unfamiliar areas. You are expected to use your learning of general principles and your judgement to answer questions based on them.

✓ Part (a) requires an estimate, so do not waste time calculating. In this example the bars are so similar that you should take the approximate average size of them and multiply by the number of relevant bars.

✓ Part (b) specifies that you are required to state two ways. You must not, therefore, give more than two answers or offer an either/or answer for either way.

Student answer

(a) 120 000 [0 marks]

(b) It has very many more working age males than expected for the size of the elderly or adults. The 30–34 year olds group is the largest adult-age group. [1 mark]

(c) The only explanation for an age-sex pyramid that is out of balance like this one is large-scale immigration of males of working age, possibly with some females. The country must be short of workers. It has a small overall population. Having few elderly, and a relatively low life expectancy suggests it is an LEDC, although it does not have a high enough proportion of young people for that and is more likely to be a newlyindustrialized country (NIC), part way to transition to an MEDC. Its relatively low number of children is more characteristic of an MEDC. [4 marks]

Total: 5 out of 7

See www.oxfordsecondary.com/esg-for-caie-igcse for the mark scheme for this question.

Examiner feedback

The answer to part (a) has been obtained by multiplying 40 000 by 3, which would be correct for either the males or the females, but is wrong for the total of both.

In part (b) the mark is gained for the second point in the mark scheme because the important words 'than expected' were written to indicate it is unusual, but the other noticeably unusual feature - a very out of balance gender divide - was not stated. It is so out of balance that a compulsory mark is allocated to it.

Part (c) is a good, well-argued answer which uses the data to support the conclusions made.

5 Population density and distribution

Key ideas

- **Population density and population distribution are influenced by a number of factors.**
- **The influencing factors are physical, economic, social and political.**

Calculation of population density

$$\text{Population density} = \frac{\text{total population}}{\text{area}}$$

Physical reasons for different population densities and distributions

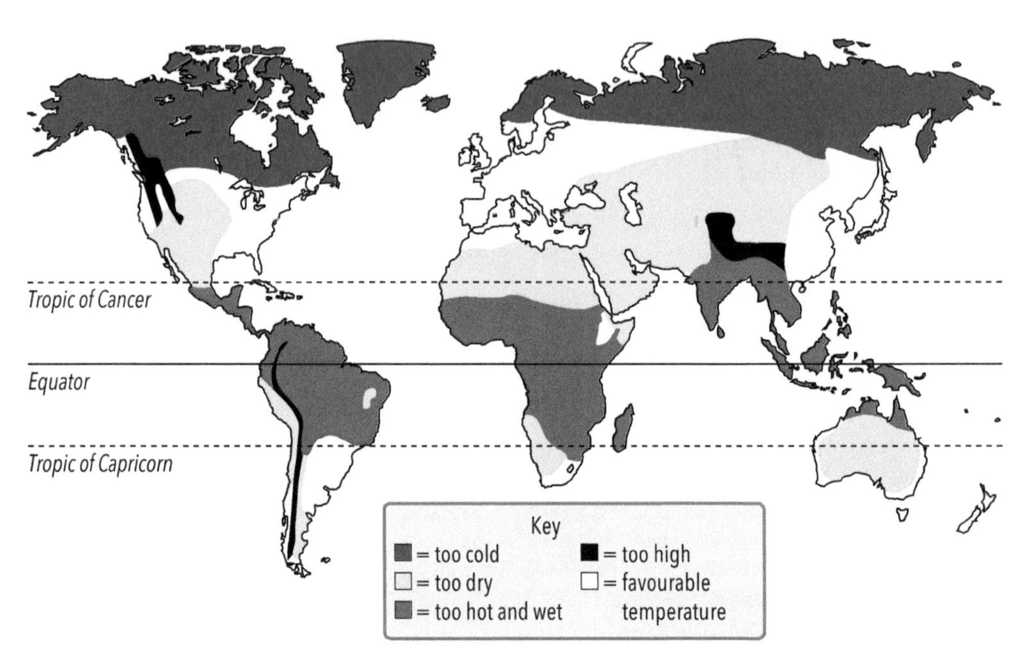

▲ **Fig. 5.1** *Map of areas with difficult conditions for population. Note that people do live in these areas, but their populations are usually sparse except in urban areas.*

Physical factor	Dense populations	Sparse populations
Relief	Low and flat lands are easy to build on and to use for economic activity.	It is difficult to build houses, transport routes and to farm on steep slopes. High lands are cold and the thin air lacks oxygen.
Soil	Fertile and thick.	Infertile, thin or stony.
Climate	Warm or hot with a rainy season or all-year rainfall, so there is a long growing season for crops and vegetation.	• Too wet – transport difficult on muddy ground. • Too dry – lacking water supplies. • Too cold for plant and crop growth. Permanently frozen ground near the surface in high latitudes.
Natural vegetation	Temperate grasslands have fertile soils.	Dense rainforests and marshy areas are difficult to penetrate.

Human reasons for different population densities and distributions

Human factor	Dense populations	Sparse populations
Economic: Farming	Areas with high crop yields from intensive farming, which requires much labour.	Areas unsuitable for farming or with large extensive farms.
Transport	Good transport for economic activity. People cluster at junctions and ports.	Areas lacking transport have little settlement or economic activity.
Industry	Industrial and mining areas with many jobs.	Lack of minerals or other resources for development. Areas without industry have few jobs available.
Social	• Some tribes and family groups traditionally live close together. partly for security. • Societies with traditional high fertility rates.	Few isolated settlements are the result of **social factors**.
Political	Some governments have built a new capital city in an undeveloped part of the country to stimulate growth.	Few isolated settlements are the result of political factors.

Common error

Be able to describe distributions of different density using east and west and north and south correctly or near a place name, e.g. 'in river valley X'. If north is at the top of the map, as it usually is, east is on the right.

Exam tip

You should be able to describe and explain a case study of a densely populated country or area and a case study of a sparsely populated country or area.

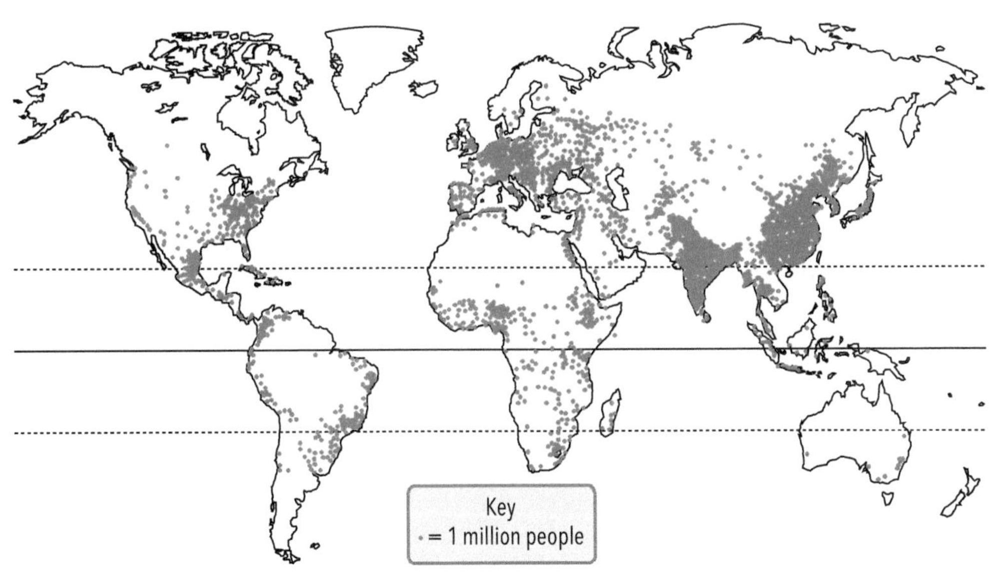

▲ **Fig. 5.2** *Dot maps show distribution well. Each dot represents the same number of people and is placed on the map where the people are.*

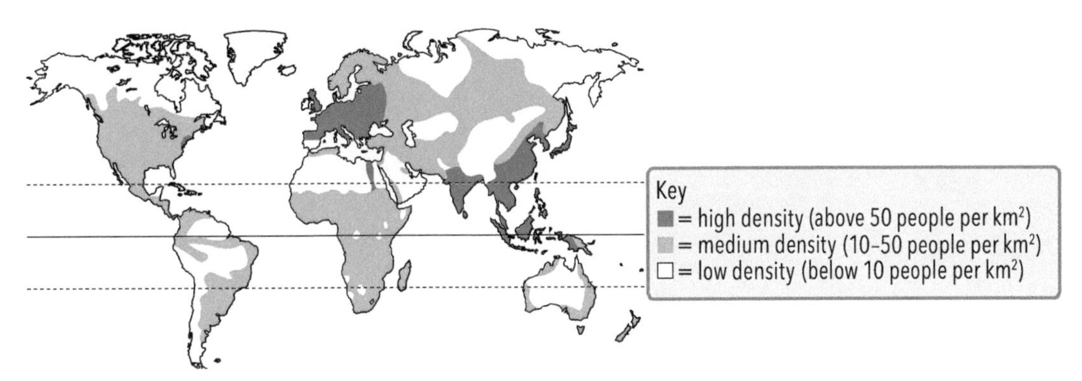

▲ **Fig. 5.3** ***Choropleth maps*** *(density shading maps) are used to show the population density of countries or separate units within countries.*

- Population density varies and is calculated by dividing the total population of an area or country by its area, usually in km^2.
- Population distribution is a description of how the population is spread over a country or area. It is densest where economic activities are not prevented by a hostile physical environment, but sparsely distributed where the natural environment is harsh.

See www.oxfordsecondary.com/esg-for-caie-igcse for the answer to the 'Apply' task.

Apply

Prepare for a question on a sparsely populated country by compiling a list of essential information. You need to be able to quote at least three relevant facts, such as the average population density of the country or sparse area, and detail about the reason for the low density (e.g. the annual rainfall).

Review

There are case studies of Bangladesh (a densely populated country) and Botswana (a country with a low population density) on pages 31–32 of *Complete Geography for Cambridge IGCSE and O Level*. The distribution of population in Canada is covered on page 33.

Pages 39–46 give information about density and distribution in rural areas and pages 48–58 about urban areas.

Pages 75–77 explain the densely populated New York area, whereas the Namib Desert study on pages 211–213 describes the difficulties of living in a hot desert.

Raise your grade

Sample question

1. For a densely populated country or area you have studied, explain why it is densely populated. **[7 marks]**

Analysis

✓ Be sure to name the country or area you have chosen.

✓ Try to mention all the physical and human factors that influence population densities.

Student answer

Bangladesh is one of the most densely populated countries in the world. It is mainly lowland that slopes very gently to the sea, with densely populated villages, particularly on the higher points of the plain. Being the delta of the Rivers Ganges and Brahmaputra, it has very fertile alluvial soils on which intensive rice cultivation supports many subsistence farmers and their families. It has a year-round growing season, being hot all year with a rainy season from which water can be stored for irrigation. The capital, Dhaka, is the most densely populated area with over 1 million people per km^2 in its squatter communities.

The country has a mainly Muslim population that has traditionally made little use of contraception, as large families have been considered desirable and helpful on family farms. Its iron ore deposits have led to some industrial development and its many waterways, the distributaries of the delta, have aided trade and port development, such as at Dhaka. Chittagong is the most important port.

[Level 3, 7 out of 7]

See www.oxfordsecondary.com/esg-for-caie-igcse for the mark scheme for this question.

Examiner feedback

This answer is brief but gains the full seven marks. Fully developed points have been given about how the country's physical and human characteristics have led to high population densities.

The example is correct. A lot of specific detail has been given: alluvial soils, river names, name of capital and port, example population density, named raw material for industry.

Sample question

2. For a sparsely populated country or area you have studied, describe and explain the distribution of population within it. **[7 marks]**

Analysis

✓ You should choose an example with at least three different areas in which the population is distributed sparsely or slightly less sparsely.

✓ You should give reasons for the variations.

Student answer

See www.oxfordsecondary.com/esg-for-caie-igcse for the mark scheme for this question.

Canada is a sparsely populated country. The most sparsely populated area is the northern half of Canada where it is extremely cold and dark in winter with frozen ground. In places, there are small communities living together to make the area slightly less sparsely populated. Some people live near the sea for fishing and other communities develop on mining camps, where minerals are found.

[Level 1, 3 out of 7]

Examiner feedback

This answer is Level 1 because it has no developed statements. It lacks explanations, such as why a cold and dark environment inhibits settlers, how frozen ground causes difficulties in the winter for mining and for transport during the summer thaw and so on.

There is nothing specific, such as the area has a population density of less than 1 person per km^2 on average. It vaguely mentions reasons for areas where population is slightly higher, but does not state anything specific about them.

The Inuit should have been mentioned in connection with the coastal settlements and a mineral and mining town named, such as gold at Yellowknife, iron ore at Schefferville or uranium at the small settlement that is called Uranium City (despite being a small town of about 5000 inhabitants).

Slightly higher population at Churchill on Hudson Bay results from its port function, exporting wheat and other products in the short summer. It is also an area often visited by tourists seeking a glimpse of a polar bear. Summers are cool and the four months or less of growing season is inadequate for crop growth.

Key ideas

- A hierarchy of settlements is a list of settlements in order of population size and the number and range of services provided.
- High-order settlements are larger, fewer in number, spaced further apart and with a wider range of services.
- Low-order settlements are smaller, more in number, more closely spaced and with a small range of services.
- A hierarchy of services puts the services of a settlement or area in rank order of importance, usually based on the population needed to support the service and the frequency of use.
- Low-order services have a small threshold population, such as a local shop or a primary school, usually in large numbers. They are found in most settlements.
- High-order services have a large threshold population, such as a department store or a university, usually in small numbers. They are usually only found in the larger settlements.
- Settlement pattern is the shape that a settlement forms on the map and how clustered or scattered the settlement is.

Hierarchy of settlements

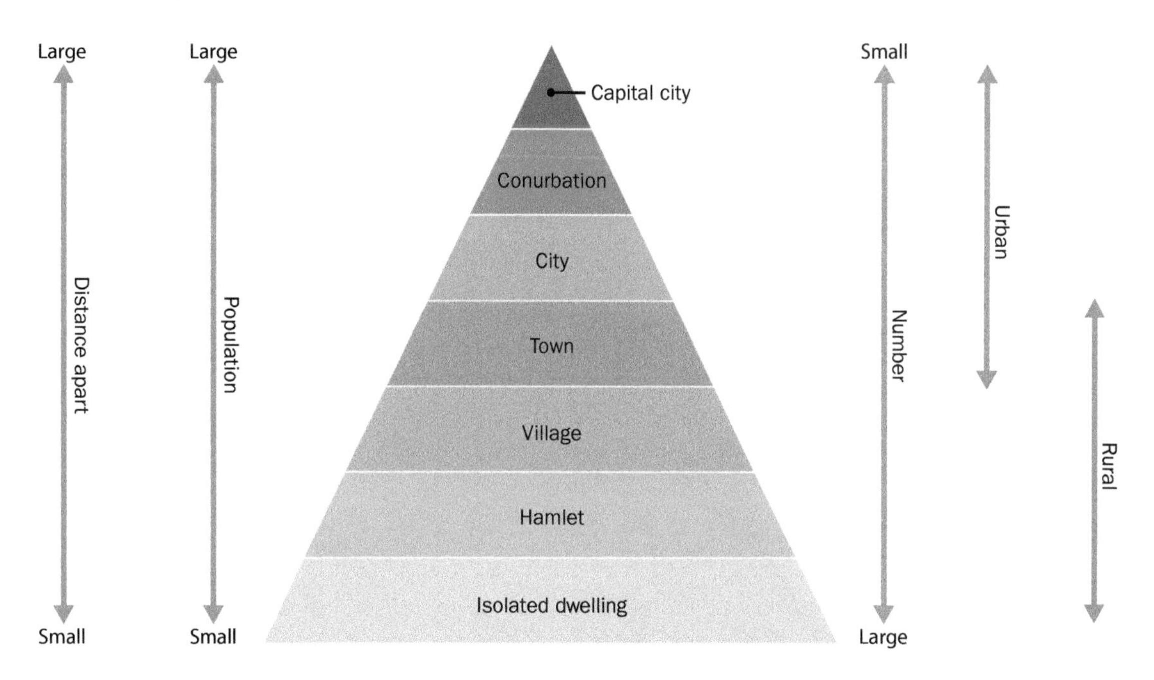

Hierarchy of services

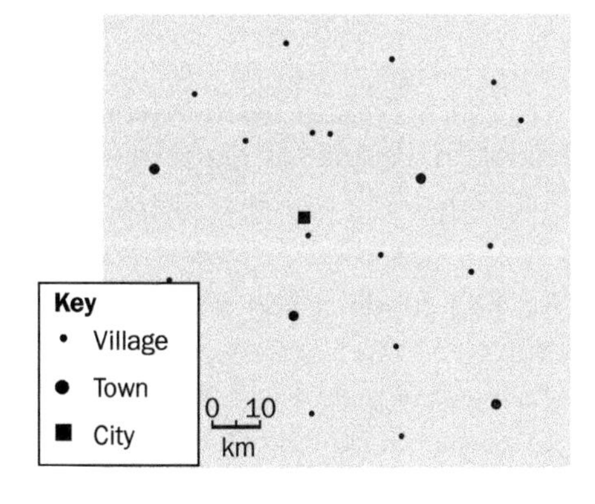

▲ **Fig. 6.1** *A map showing the settlement hierarchy in an area. Which settlement is highest in the hierarchy?*

Exam tip

Make sure that you can describe the settlement hierarchy of an area that you have studied, including the **services** in the settlements.

Common error

Do not confuse the hierarchy of ***settlements*** with the hierarchy of ***services***.

Sphere of influence

This is the area served by a **settlement** or **service**.

The size of the sphere of influence depends on:

- the size of the settlement and the services provided
- the population density of the area (which may in turn be affected by factors, such as relief)
- the wealth of the people in the area
- transport facilities
- competition from other settlements.

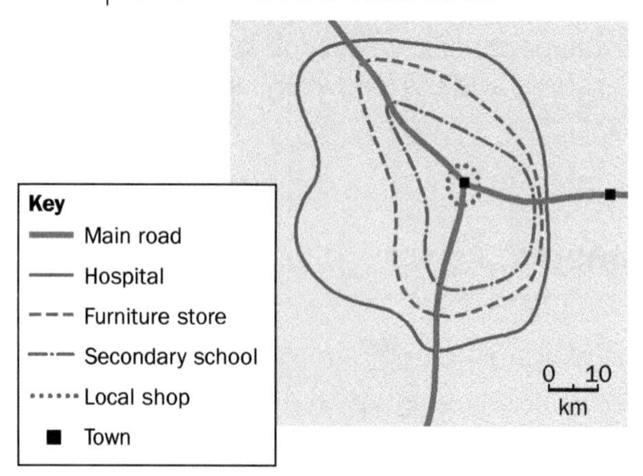

▲ **Fig. 6.2** *A map showing the spheres on influence of different services. Which service is used most frequently?*

Settlement pattern

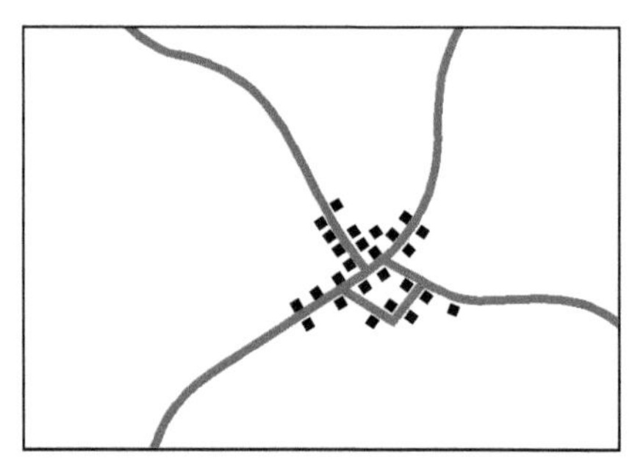

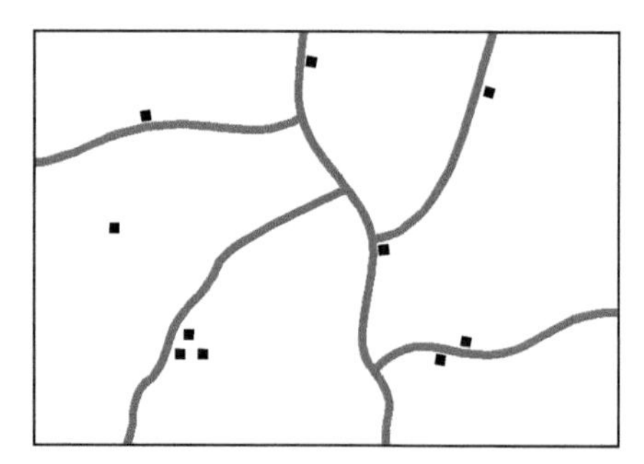

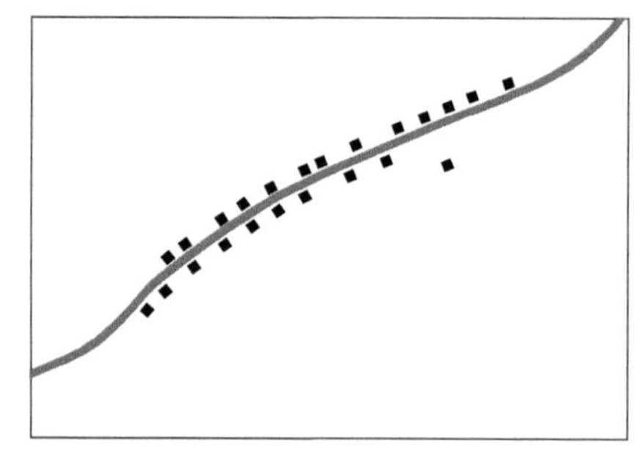

▲ **Fig. 6.3** *Maps showing settlement patterns: nucleated, dispersed and linear. Can you see which is which?*

Exam tip

Make sure that you can describe how each of these factors will affect the size of the sphere of influence.

- **Nucleated** settlements have houses clustered together as villages, with fewer isolated dwellings. The shape of the villages is compact and more square or circular. They develop at cross roads, bridges, good defensive points and where there are mineral resources. People can enjoy the social benefits of living close to their neighbours. They have easy access to services like shops and schools.
- **Dispersed** settlements are scattered, isolated dwellings and small hamlets. They develop where the agricultural land is poor and where people need large areas of land for things such as grazing. It would be impossible to live in a village and still be within travelling distance of agricultural land.
- **Linear** settlements are in long thin rows, often along roads or tracks. They develop along a road or track for transport, along a river or a line of springs for water supply or along a valley floor avoiding steep valley sides.

Recap

Settlement pattern is the shape that a settlement forms on the map and how clustered or scattered the settlement is. There are three main types: nucleated (clustered); linear (in a line); and dispersed (scattered).

Apply

For each of the settlements A, B, and C shown below, describe the settlement pattern and give reasons for its location.

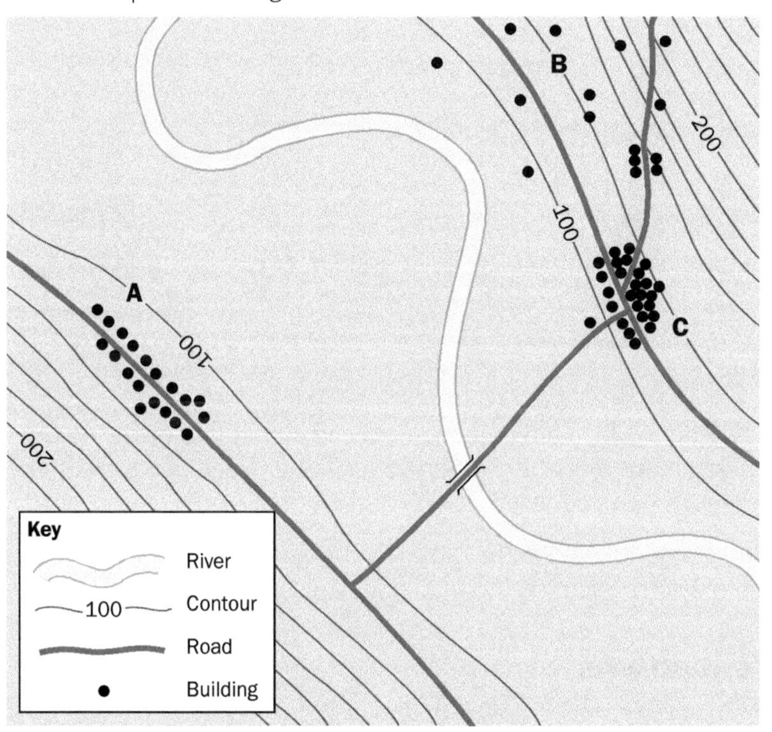

See www.oxfordsecondary.com/esg-for-caie-igcse for the answers to the 'Apply' task.

Review

Nucleated settlements

- at cross roads
- bridges
- good defensive points
- mineral resources

Dispersed settlements

- poor agricultural land

Linear settlements

- along roads or tracks
- along a river or a line of springs for water supply
- along a valley floor avoiding steep valley sides

Sample question

1. Study the map and answer the following questions.

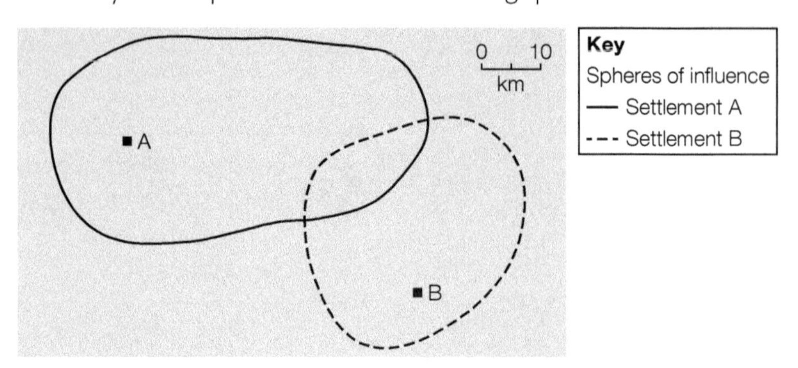

(a) What is the term used to describe the greatest distance that people travel to access a service? **[1 mark]**

(b) Suggest why the spheres of influence overlap. **[2 marks]**

(c) Suggest reasons why settlement A has a larger sphere of influence than settlement B. **[4 marks]**

Analysis

✓ In questions which are point marked, remember that the more points that you make the more marks you are likely to get. Therefore, try to suggest at least two reasons in (b) and four reasons in (c).

✓ The command word in the question is 'suggest', which means you can give answers which might be true but you can't be certain.

Student answer

See www.oxfordsecondary.com/esg-for-caie-igcse for the mark scheme for this question.

(a) Range [1 mark]

(b) People don't always go to the nearest settlement for a service. [1 mark]

(c) This could be because some settlements have more and better services. It could also be that some areas are sparsely populated, so a greater area is needed to find the threshold population. [0 marks]

Total: 2 out of 7

Examiner feedback

In part (b) the candidate has made one good point and scores one mark. The mark scheme lists three other possible points. In part (c), the candidate refers to more or better services and to population density, both of which are correct ideas. However, they should have linked these ideas to settlements A and B, e.g. A has more services and is in a sparsely populated area.

Key ideas

- The physical geography of an area (e.g. relief, soils, water supply, drainage) influences the development of rural settlements.
- Some settlements develop into larger towns because they have particular advantages.
- Rural settlements in many parts of the world face problems, such as depopulation.

Factors influencing the sites and development of rural settlements

You should be able to describe the effect of the following factors:

- agricultural land use
- relief: altitude, gradient, **aspect**
- soils
- water supply
- drainage and flooding
- river crossings
- natural resources, such as minerals
- development of tourism
- **accessibility**.

These factors influence the **site**, **situation** and pattern of settlement on the following map and cross-section.

Common error

Make sure that you know the difference between site and situation.

Exam tip

Make sure that you can describe the distribution of settlement in an area that you have studied.

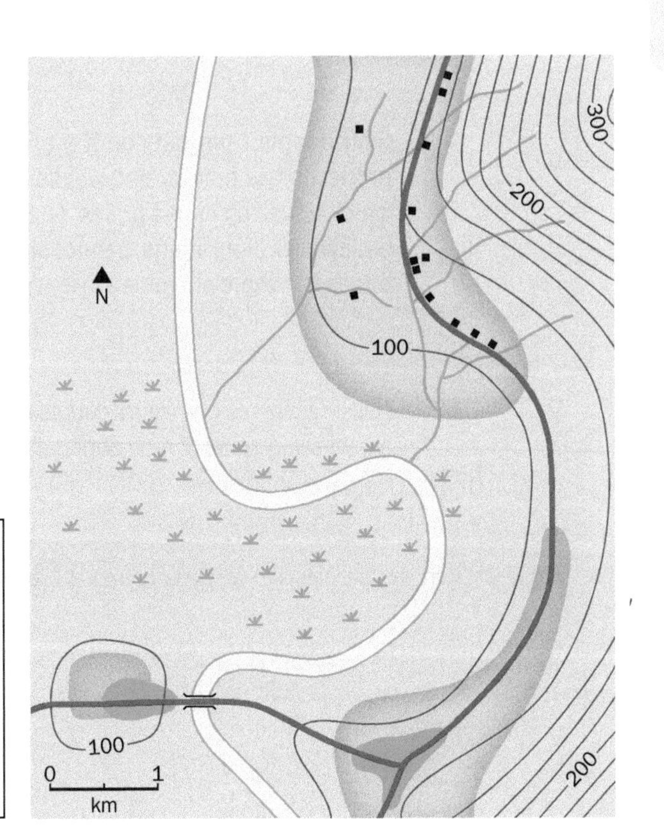

▲ **Fig. 7.1** *The location of settlements in a rural area. Give reasons for the location of the villages and dwellings.*

Factors influencing the size, growth and functions of urban settlements

You should be able to describe the following factors which have allowed rural settlements to grow into larger towns and cities:

- **route centres** (nodal points)
- agricultural centres
- ports
- administrative towns and cities.

Problems and changes in rural settlements

1. Because of low wages and a lack of jobs, young people leave their villages to work in urban areas.

↓

2. The average age of the remaining population increases.

↓

3. Services, such as schools, post offices, public transport, and shops, begin to close, because there are fewer people to use them.

↓

4. Rich people from outside the area buy up properties as holiday and weekend homes, which drives up house prices. Local people, especially young adults, cannot afford these prices and more are forced to leave the area.

↓

5. This encourages even more people to leave, and a downward spiral of rural depopulation occurs.

▲ **Fig. 7.2** *Common problems in villages in remoter areas of MEDCs*

Common error

Remember that rural-urban migration can be a problem in MEDCs as well as LEDCs.

Exam tip

Expect to have to describe settlement patterns on maps of areas which are unfamiliar to you.

1. For a rural area or a village you have studied, explain the factors that have affected its growth.
2. The following list describes some of the problems that can affect a rural area. Put them in the order in which they might happen.

- Services such as schools, post offices and public transport close
- Rich people buy property to use as holiday homes
- Young people leave the area
- Average age of the population increases
- Lack of jobs
- House prices rise

- The location of a settlement is not determined by chance. The site and situation will have particular advantages which will have encouraged its growth.
- Some rural settlements do not continue to grow and people begin to leave and services go into decline.

See www.oxfordsecondary.com/esg-for-caie-igcse for the answers to the 'Apply' task.

Factors encouraging the location of settlements.

Agricultural land use and soils	**Rich agricultural land is usually more densely settled.**
Relief: altitude, gradient, aspect	Lowland areas are warmer and more densely settled. Gentle slopes are easier to build on and plough. Sunny slopes (south facing in the Northern Hemisphere; north facing in the Southern Hemisphere) are warmer.
Water supply	Encourages settlement at **wet point** sites on rivers and springs.
Drainage and flooding	Settlements often avoid flood plains and are at **dry point** sites.
Natural resources (such as minerals)	Settlements grow up around mines.
Development of tourism	Resorts develop near beaches or at areas of scenic beauty.
Accessibility	Linear settlements develop along roads, but nucleated settlement develop at junctions and bridge points.

Factors which encourage rural settlements to grow into larger towns and cities.

Nodal points (route centres)	**As road junctions are so accessible, they often develop as service centres for the surrounding area.**
Agricultural centres	Rich agricultural areas require towns to provide services and to market or export their produce.
Ports	Ports grow where there are good sheltered, natural harbours and good links with a **hinterland** from which goods can be exported.
Administrative towns and cities	These are the towns and cities which have grown into regional or national capitals.

These factors are described on pages 40–42 and pages 52–54 of *Complete Geography for Cambridge IGCSE and O Level.*

Sample question

1. The map shows four settlements: A, B, C and D.

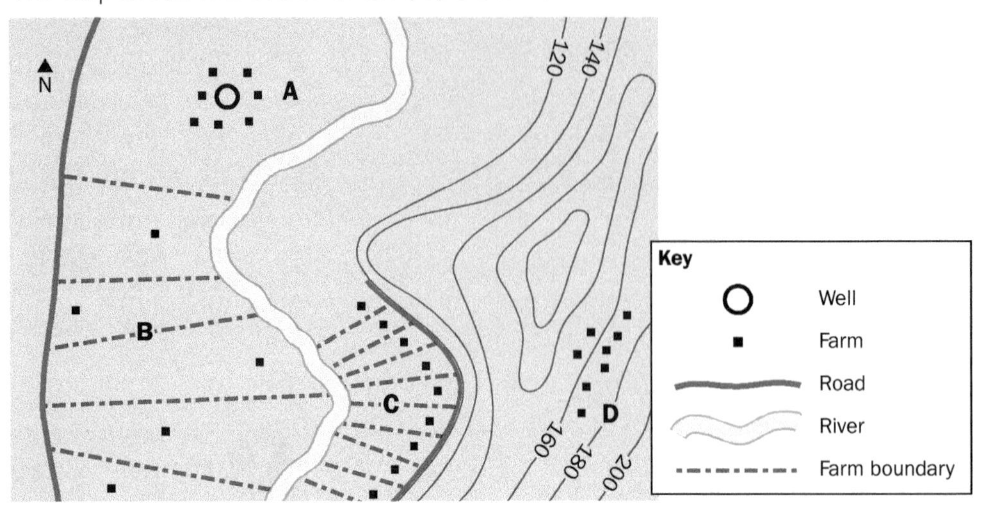

(a) Identify the settlement pattern in settlement B. **[1 mark]**

(b) Give **one** advantage and **one** disadvantage of the site of settlement A. **[2 marks]**

(c) Using evidence from the map, suggest reasons for the growth of settlements C and D. **[4 marks]**

See www.oxfordsecondary.com/esg-for-caie-igcse for the mark scheme for this question.

Analysis

✓ This question is completely based on the map. You use your knowledge to help answer the question, but all the information comes from the map.

✓ In part (c), remember that you should always give map evidence for your reasons.

Student answer

(a) Dispersed [1 mark]

(b) Settlement A is on the floodplain therefore it has the disadvantage that it might flood. However, the farms are clustered around the well, which might help with irrigation in dry weather. [2 marks]

(c) Settlement C is a linear settlement which has been built along the road for access. The farms have their own plots of land leading down to the river for water supply. Settlement D is also linear, but does not have any road access. It is on gentle sloping land which is easier to build on and is above the floodplain, therefore is not liable to flooding. [4 marks]

Total: 7 out of 7

Examiner feedback

This is an excellent answer. In part (c) evidence has been given from the map and each point is fully explained.

8 Urban settlements

Key ideas

- There is a range of different types of land use in towns and cities.
- Urban morphology is the distribution of different types of land use in a town or city.
- Models of urban morphology help to explain patterns of land use in towns and cities.
- A range of problems occur in urban areas in MEDCs and LEDCs.

Urban models

You should be able to describe:

- concentric zone model
- sector model
- a model of cities in LEDCs.

Common error

Do not confuse the models for MEDC cities with the model for LEDC cities, which is very different.

Types of urban land use

You should be able to describe each of the following land use types:

- central business district (CBD)
- high-density housing
- low-density housing
- apartments (flats)
- shanty houses
- open spaces
- transport routes
- industrial areas
- rural-urban fringe.

Exam tip

Make sure that you can describe the urban morphology of a town or city that you have studied and explain how it has developed.

Urban problems

Problems	Solutions
Decline of the CBD in cities in MEDCs Issues include: • congestion and poor accessibility • lack of parking space • high land prices • retailers leaving (smaller shops which can no longer afford the high rents; major department stores and hypermarkets have moved to greenfield sites on the outskirts), leading to empty buildings • decentralization of companies and administration to the outskirts.	**Pedestrianization** • Allow a more safe, relaxed environment. • Less air and noise pollution from vehicles. **Shopping malls** • Undercover shopping areas. • Air-conditioned malls. • Cafés and small restaurants. **Visual improvements** • Flower beds, seated areas, trees and hanging baskets. • Pavement cafés and bars are introduced. **Transport improvements** • Underground railways.

Common error

When the term inner city is used for MEDC cities, it means the inner residential zone, not the CBD.

Exam tip

Make sure that you can explain how historical development, route intersections and land prices can explain the development of the CBD.

Problems	Solutions
Security in the CBD, especially in evenings, in cities in MEDCs and LEDCs • High crime rates • Litter • Graffiti	• Patrols by police or by private security firms. • Closed circuit TV is a deterrent to pick pockets and shoplifters.
The twilight zone in cities in MEDCs This is the transition zone on the edge of the CBD. It can be an area of decline and can suffer from: • derelict land and buildings • high rates of crime and social problems.	Re-development by city governments.
Crime and racial conflict in cities in MEDCs and LEDCs • High levels of poverty. • Development of **ghettos** in inner cities in MEDCs. • Informal settlements in LEDCs (see Chapter 9).	• Providing social facilities, such as sports clubs. • Job creation schemes to provide employment. • Special projects that bring communities together. • Zero tolerance on crime. • Ensuring adequate policing on the streets. • Providing language lessons for immigrants.
Housing shortages in MEDCs • Older properties nearer to city centres requiring renovation or renewal. • A population increase through immigration and natural growth. • Property prices being too high for those who are unemployed or on low wages.	• **Slum** clearance schemes. • Older housing replaced by blocks of flats or new houses in the suburbs. • New towns built in the countryside. • In Japan, land reclaimed from the sea to build houses on.
Traffic congestion in cities in MEDCs and LEDCs • Ancient cities built long before the need for mass public transport. • Increased use of private cars. • Large numbers of commuters so that many of the trains and buses that carry these people are not needed during the rest of the day. • People visiting the CBD for sightseeing, shopping or entertainment. • People passing through the city on their way to other places.	• Underground railways • Bus lanes • Congestion charge • Electronic ticketing • Integrated transport policies • Ring roads • Traffic lights (robots) • Tidal flow • Trams • Roundabouts (circles) • Park and ride schemes

Problems	Solutions
Air pollution • Urban areas have less clean air than the surrounding countryside. • It is in the major cities of LEDCs and NICs that the highest levels of air pollution occur.	• Laws to control the emissions from industry and housing, including smoke-free zones. • Carrying out checks on vehicle exhausts and removing polluting vehicles. • Higher taxes for the most polluting vehicles. • Introducing lead-free petrol. • Reducing the amount of electricity generated from thermal power stations. • Developing new power stations that do not release carbon dioxide into the air.
Water pollution • Raw sewage in rivers and groundwater. • Contamination of drinking water leading to health issues, such as diarrhoea and dysentery. • This is a particular problem in some LEDCs.	• Provision of proper swage systems and clean piped water.
Visual pollution • Ugly buildings and industry • Derelict land • Graffiti • Litter	• Stricter planning regulations. • Improve refuse collection.
Noise pollution • Cars and lorries, trains, aircraft taking off and landing. • Factories, large congregations of people, such as football crowds. • Noise in residential areas, for example from radios and parties.	• Laws which limit the noise from factories and homes. • Separating noisy areas from residential areas. • Building solid fences along motorways and major roads to reduce the noise reaching residents.
Urban sprawl • High car dependence. • Inadequate services within the spreading suburbs, such as shops, doctors. • High cost of providing social facilities. • High costs for public transport. • Lost time and productivity for **commuting**. • High levels of racial and socio-economic segregation. • Changing character of the countryside and loss of the rural way of life. • Rural settlements turned into **dormitory villages** for long distance commuters.	• Careful urban planning, such as green belts. • Provision of social facilities. • Building new towns beyond the green belts.

Common error

Some of these problems only occur in MEDCs, some only in LEDCs and some in both.

Exam tip

Make sure that you can describe some of these problems and solutions in a town or city that you have studied.

Recap

- All the urban models have a central business district (CBD).
- In the concentric zone model, the CBD is surrounded by a ring of light manufacturing, then a ring of low-class residential, then a ring of middle-class residential and finally a ring of high-class residential zones.
- In the sector model there are some concentric zones, but most of the zones radiate from the centre, sometimes along radial routes.
- In models for LEDC cities, the residential zones are the opposite to the concentric zone model: the better housing is closer to the CBD and recent squatter housing is on the outskirts.

See www.oxfordsecondary.com/esg-for-caie-igcse for the answers to the below 'Apply' task.

Apply

1. Study the diagram which shows the quality of life in cities in LEDCs and MEDCs.
   (a) Describe the differences in quality of life between cities in LEDCs and cities in MEDCs.
   (b) Give reasons for the differences which you have described.
2. Study the diagram which shows how the height of buildings has changed in Shanghai (China) between 1980 and 2005.
   (a) Describe the heights of the buildings in Shanghai in 1980.
   (b) How has this changed by 2005?
   (c) Suggest in which zone of the city the buildings of over 30 storeys will be found and explain your choice.

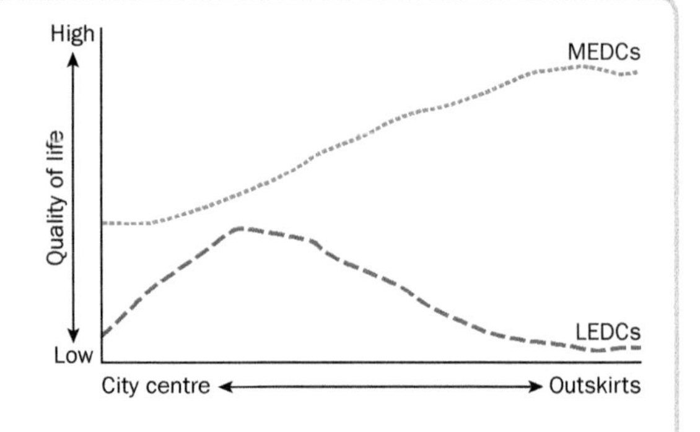

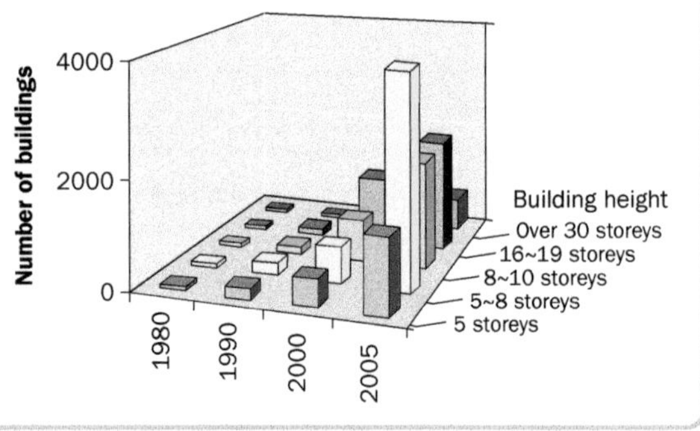

Review

- The CBD includes government buildings, high-order retail services, offices, theatres, hotels, restaurants, multi-storey buildings, public transport services (including underground railways), few residents, high numbers of pedestrians during the day.
- The CBD has developed because: it was the original growth point of the city, it was the point where roads from the outskirts converge, which made this the most accessible area of the town, land prices became high and so buildings were taller.
- In shanty/squatter/informal settlements, residents do not have a legal right to occupy the land. The houses are not weatherproof and can be cold in winter. There may be no proper sanitation, electricity and water supply, leading to diseases. There may be no refuse collection. There is extreme poverty, overcrowding and high unemployment. There are high levels of crime, as well as drug and alcohol abuse. The residents may be recent immigrants (see Chapter 9).
- Solutions to the problems of squatter settlements include: low-cost housing schemes, which provide basic housing with running water, electricity and proper sanitation; self-help schemes, which provide groups of people with the materials to build proper houses, provision of basic services (such as sewerage, piped water and refuse collection) by the city authorities.

Sample question

1. For a town or city you have studied, describe the problems of the CBD. **[7 marks]**

Analysis

✓ The answer should be all about the CBD and not other urban problems.

✓ To get to Levels 2 and 3, each point made should be fully explained. The answer need not be very long – it is better to make a small number of points and explain them thoroughly.

✓ The answer should contain specific information about the chosen example.

Student answer

My example is Barrow-in-Furness, UK. The CBD is a zone of change. Problems arise because land there is expensive. There are shops boarded up and properties in a poor state of repair.

Increased car ownership has led to congestion and a lack of parking space. Retailers have left the CBD. Small shops can no longer afford the high rents. Major department stores and hypermarkets have moved to greenfield sites on the outskirts of town. Here the land is cheaper and people with cars can also drive easily to these locations, and there is plenty of parking space.

The growth in online shopping, particularly for clothing has also forced stores to close. Some people find the CBD a rundown area with empty derelict buildings, which is unsafe and dirty and suffers from litter and graffiti.

Some offices have relocated to new purpose-built premises on the outskirts of the town, where land prices are lower and employees can drive to work and park more easily.

In the evening, the CBD can be very empty. The only services open are bars, restaurants and nightclubs.

On the outskirts of the CBD there is derelict land and buildings and some housing, with high rates of crime and social problems, such as drug addiction. [Level 2, 6 out of 7]

Examiner feedback

The answer contains several correct statements, each of which is fully developed and explained well. There is a relevant example, but there are no specific details of the example, therefore the answer does not reach Level 3. The student could have achieved six marks even with fewer statements.

See www.oxfordsecondary.com/esg-for-caie-igcse for the mark scheme for this question.

Sample question

2. For an example of a city you have studied, describe the problems caused by rapid urban growth. **[7 marks]**

Analysis

✓ The answer could be for an LEDC city and focus on squatter settlements or it could be for an MEDC city and focus on urban sprawl.

Student answer

See **www.oxfordsecondary.com/esg-for-caie-igcse** for the mark scheme for this question.

Khayelitsha is 15 to 20 km from the centre of Cape Town, South Africa. In 2011, around 62% of residents in Khayelitsha were rural–urban migrants, most coming from the Eastern Cape and settling in the outskirts of Cape Town in search of work.

70% of residents still live in shacks and one in three people has to walk 200 metres or further to access water. Household parts are for sale along the road, including wooden frames, corrugated zinc and mattresses, which are all extremely vulnerable to fire.

Only 53% of the working-age population is employed and 89% of households are either moderately or severely food insecure.

Crime rates are high, with communities split along racial lines. There is an AIDS pandemic and crime is often drug-related.

Khayelitsha has a very young population – fewer than 7% of its residents are over 50 years old and over 40% of its residents are under 19 years of age. This means that growth is continuing. [Level 3, 7 out of 7]

Examiner feedback

This is a good answer, which gives a lot of information specific to the chosen example. The answer covers several problems including the poor quality buildings, water supply, fire risk, unemployment, food supply and crime. Each point is described using specific detail about the chosen city, for example statistics.

9 Urbanization

Key ideas

- **Urbanization has occurred at different rates in different continents and levels of urbanization vary greatly.**
- **The rate of urbanization has increased greatly.**
- **Migration to cities has created problems in the areas people have left and in the areas that they are moving to.**
- **Shanty settlements have unique problems which are proving difficult to solve.**

Rates of urbanization

Continent	Level of urbanization	Current rate of urbanization
Richer countries of Europe and North America, and other MEDCs	More than 90% (started over 200 years ago when these countries went through the Industrial Revolution)	Slow or stopped and **counter-urbanization** is occurring
South America	Often over 75%	Slowing
LEDCs of Africa and south-east Asia	May be less than 50%	Rapid

Common error

Urbanization is not just about the growth of towns and cities, it is the increase in the *percentage of the population* living in towns and cities.

Causes of urbanization

- Overall population growth
- Rural–urban migration
- Increasing numbers of people working in secondary and tertiary industries, which are concentrated in urban areas. This happened earlier in MEDCs.

Exam tip

There is overlap between this topic and population migration. It may come up on the paper in a population question or in a settlement question.

Impact in rural areas

- Rural depopulation has occurred.
- This has led to a shortage of labour, particularly in agriculture, as the migrants are usually of working age.
- The population has large numbers of young and elderly dependents.

Common error

Make sure you know if a question is asking about all dependents or just young or old dependents.

Shanty settlements

Possible problems
Residents do not own the land or have a legal right to occupy the land and could be evicted at any time.
Shacks are not properly suitable for habitation and pose a risk to life and health. • They are prone to fires. • They are not weatherproof and can be cold in winter. • There is no proper sanitation, leading to diseases such as cholera. • People often have to walk to the nearest water supply. • There is no refuse collection. • There is no electricity supply, or if there is, it is illegal. • There is frequently extreme overcrowding, with families living in one or two rooms.
They are not situated in locations close to work. • They are often on the outskirts of cities. • Inhabitants must travel long distances to the centre of the city for work and there is little public transport. • Extreme poverty and high unemployment.
They are often dangerous areas. • High levels of crime, and drug and alcohol abuse. • Racial tensions between migrant groups.

Exam tip

Make sure that you know a case study that illustrates the impact of rural–urban migration in an LEDC, both on the rural area people are leaving and on the urban area they are going to, including **shanty** settlements.

Exam tip

Remember that for case study questions, you need to identify your example as precisely as possible and give specific details about it, not just write in general terms about the topic.

Possible solutions

- The government could provide low-cost housing schemes.
- Self-help schemes could provide groups of people with the materials to build proper houses.
- The properties would have piped water, electricity, sewers and refuse collections, making them less of a risk to health.

Urbanization	When?	Why?
Europe and North America	Started 200 years ago. Now stopped.	Industrial Revolution. People moved to towns to work in factories.
South America	Since 1950. Slowing.	Overall population growth. Rural–urban migration. Increasing numbers of people working in secondary and tertiary industries, which are concentrated in urban areas.
Africa and south-east Asia	Since 1950. Rapid.	

Urbanization is associated with rural–urban migration, which has created problems in the rural areas people have left, and in the urban areas they have moved to.

See www.oxfordsecondary.com/esg-for-caie-igcse for the answer to the 'Apply' task.

Choose an example of rural urban migration. Make a list of its impacts in the rural area that people have left and in the urban area that they have moved to.

Impacts in the rural area

- Rural depopulation has occurred.
- Shortage of labour, particularly in agriculture, as the migrants are usually of working age.
- The population has large numbers of young and elderly dependents.
- Development has been hindered.

Impacts in the urban area

- Shanty housing has developed (see Chapter 8 for details).
- Extreme poverty and high unemployment.
- Extreme overcrowding with families living in one or two rooms.
- Racial tensions between migrant groups.

Raise your grade

Sample question

1. Study the graphs of population in rural and urban areas and answer the questions that follow.

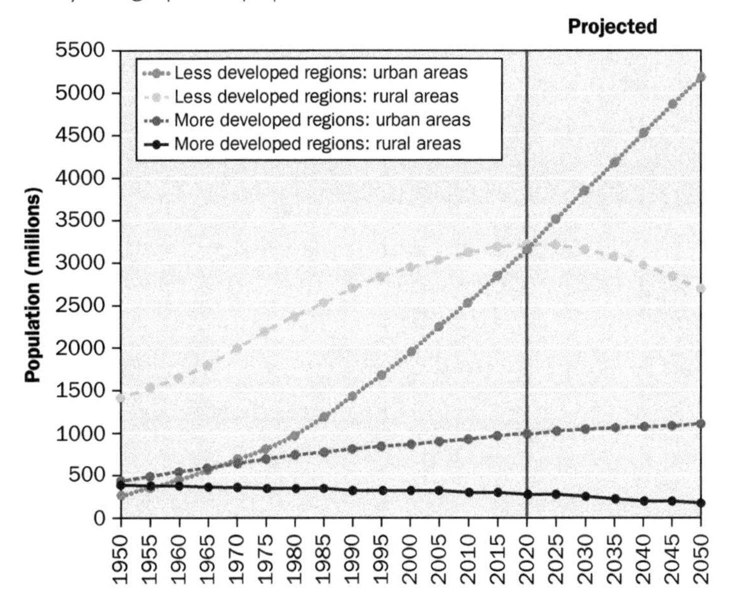

(a) Explain the meaning of the term *urbanization*. **[1 mark]**

(b) Using the graphs, in 2000, which of the four regions had the highest and the lowest populations? **[2 marks]**

(c) Explain the causes of the changes shown on the graphs. **[4 marks]**

Analysis

✓ Part (b) can be answered from the graphs, but in part (c) you have to use your own knowledge to give reasons for what the graphs show.

✓ Note that the command word 'explain' is used here, which means you must give reasons why something is happening.

Student answer

(a) The growth of cities. [0 marks]

(b) Less developed urban areas have the greatest growth and more developed rural areas have the lowest growth – actually showing a decline. [2 marks]

(c) In MEDCs, birth and death rates are lower and more equal so that the populations have been more stable. Growth is greater in urban areas due to migration and ageing populations in rural areas. In LEDCs, there has been an overall population growth because birth rates exceed death rates, which are decreasing. At the same time, rural–urban migration has occurred and is expected to continue. [4 marks]

Total: 6 out of 7

Examiner feedback

The candidate has lost just one mark for the definition of urbanization in part (a). Compare the candidate's definition with the one in the glossary at the back of this book. The answer to (c) is excellent.

See www.oxfordsecondary.com/esg-for-caie-igcse for the mark scheme for this question.

The natural environment

Chapters 10 to 19 are designed to help you know and understand the main features of the natural, geographical environment, such as volcanoes, earthquakes, rivers, coastlines, weather, climate and vegetation. There is emphasis on the natural processes but also on the links between the natural environment and human activity.

There will be two questions on *The natural environment* in the *Cambridge IGCSE® or O Level* Paper 1. There are also often two questions in Paper 2. Paper 4 frequently has questions on rivers or coasts. Volcanoes, earthquakes, rivers, coasts and weather are usually tested as separate topics. Questions on tropical rainfall and hot deserts often stress the links between climate and vegetation.

Key ideas

- The Earth's surface is made up of a series of sections known as plates.
- The plates are, on average, about 50 kilometres thick and include the Earth's crust and the upper part of the layer below, which is called the mantle.
- The plates are relatively cold and rigid. The rocks underneath the plates have temperatures of more than 1300 °C and behave plastically.
- The plates can move relative to each other, flowing over the hotter, more plastic rocks below, which act like a lubricated layer.

Plate boundaries (plate margins)

Type of plate boundary	Examples	Type of stress	Features
Destructive with an oceanic plate and a continental plate	Andes	Compression	• Earthquakes • Fold mountains • Volcanoes • Ocean trenches • **Subduction**
Destructive with two oceanic plates	Japan Philippines West Indies	Compression	• Earthquakes • Island arcs • Volcanoes • Ocean trenches • Subduction
Collision of two continental plates	Himalayas	Compression	• Earthquakes • Fold mountains
Constructive	Mid-Atlantic Ridge East Pacific Rise Carlsberg Ridge	Tension	• Earthquakes • Ocean ridges • Volcanoes
Conservative	San Andreas **Fault**	Shearing	• Earthquakes

Common error

Constructive (divergent) and destructive (convergent) boundaries are often confused. Make sure you know which is which.

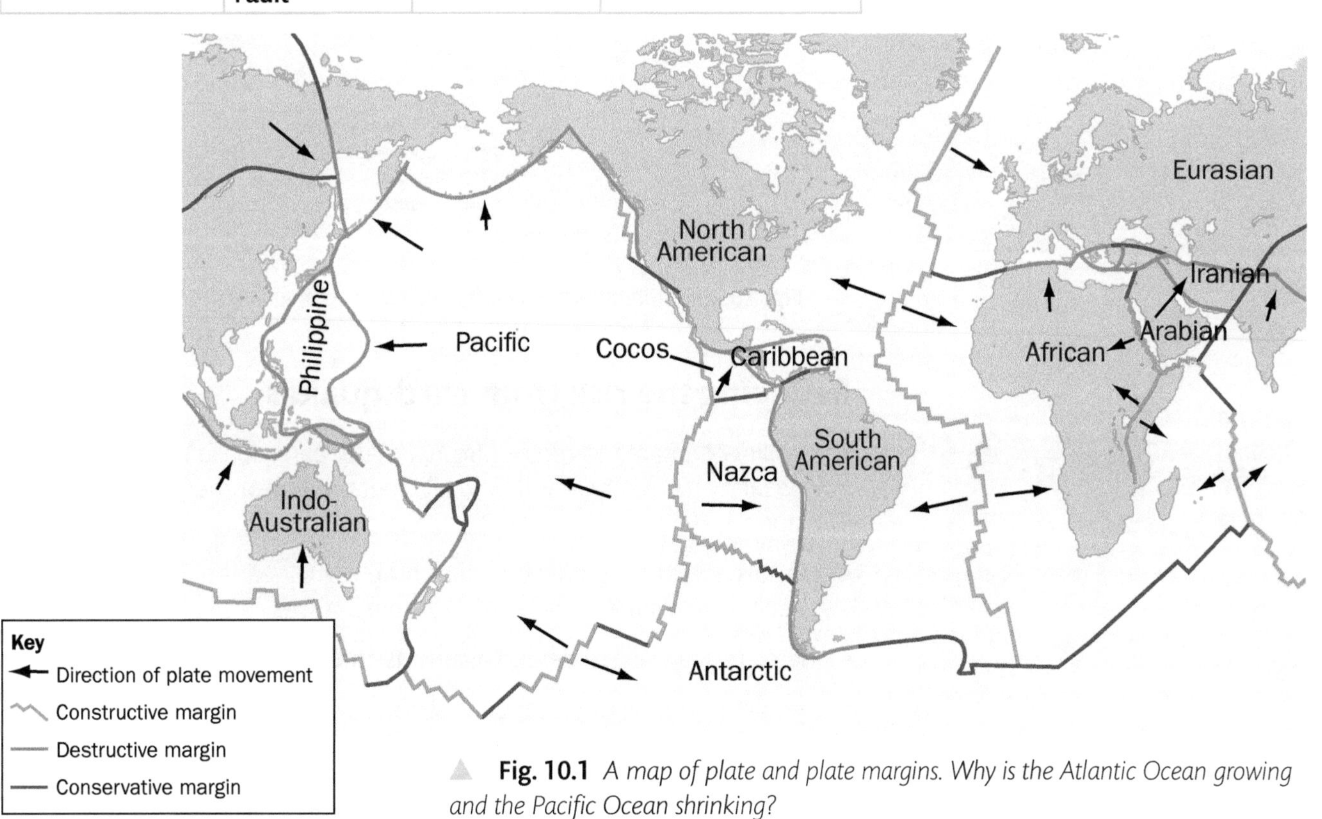

Fig. 10.1 *A map of plate and plate margins. Why is the Atlantic Ocean growing and the Pacific Ocean shrinking?*

Common error

In exams, candidates sometime write about volcanoes when a question is about earthquakes.

Exam tip

Make sure that you know a case study of a major earthquake and can describe its causes and effects.

Earthquake damage

The effects of an earthquake are described on the **Mercalli Scale**. The amount of damage an earthquakes causes will be affected by the following factors:

- energy released (**Richter Scale**)
- depth of the **focus** beneath the surface (shallower earthquakes have a greater effect)
- density of population
- whether or not the buildings have been built to withstand earthquakes
- how solid the bedrock is; weak sands and clays can turn to liquid, known as liquefaction, causing buildings to collapse.

Fig. 10.2 *Earthquakes can cause devastation in a variety of ways*

Reducing the risk from earthquakes

The ability of an area to recover from a major earthquake is affected by how wealthy a country is and how efficient the government is. In MEDCs like Japan:

- schools and other public buildings have regular earthquake drills, so that people are prepared when an earthquake strikes
- there are sea walls to defend against **tsunami**
- they are part of an international tsunami warning system
- buildings are constructed to withstand earthquakes. Can you list some of these building design features?

Recap

Earthquakes occur at all types of plate margin and are a result of movements of the plates. The point of origin of an earthquake is its focus. The plates try to move but become stuck. They suddenly fracture and move, releasing the built-up energy.

Apply

Name features A to E on the following cross-section diagram. What type of plate margin is shown?

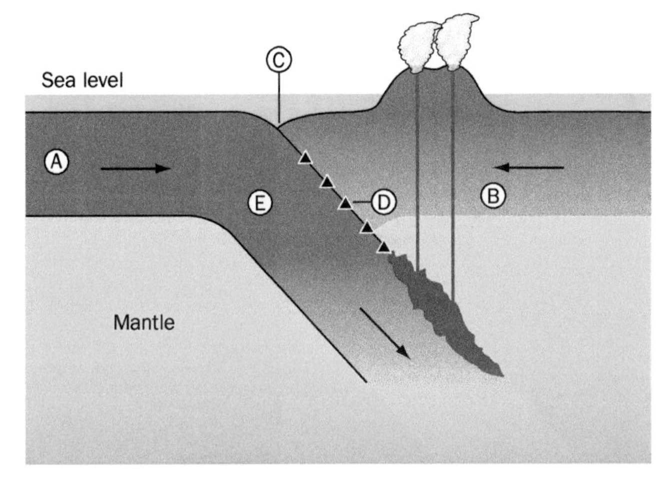

See www.oxfordsecondary.com/esg-for-caie-igcse for the answer to this 'Apply' task.

Review

- The plate margin shown is a destructive (convergent) margin with a continental plate and an oceanic plate.
- Notice how the foci increase in depth further down the subduction zone.
- An example of this situation is the Andes mountains in South America.

Sample question

1. Look at the map of the world showing the locations of earthquake epicentres and answer the following questions.

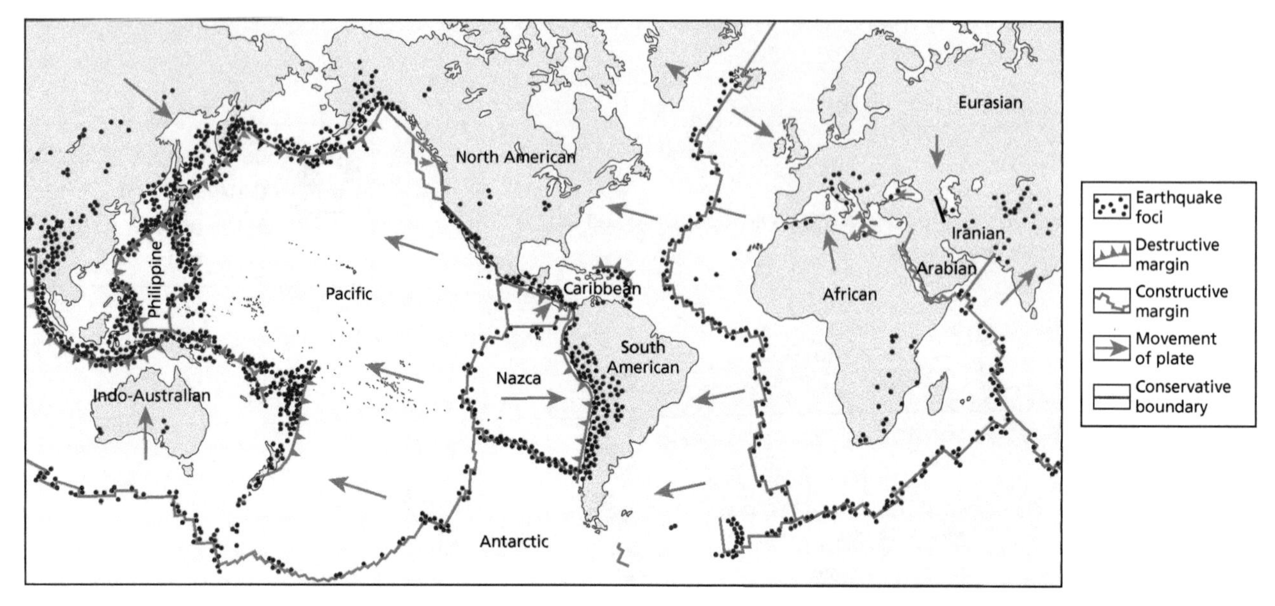

(a) Define the term earthquake *epicentre*. **[1 mark]**

(b) Identify **two** other features found in these zones. **[2 marks]**

(c) Describe the distribution of earthquakes shown on the map. **[3 marks]**

Analysis

✓ Parts (a) and (b) require you to use your own knowledge.

✓ Part (c) can be answered using the map alone. Notice that the command word is 'describe' and not 'explain'.

See www.oxfordsecondary.com/esg-for-caie-igcse for the mark scheme for this question.

Student answer

(a) This is the point on the Earth's surface directly above the focus. [1 mark]

(b) Volcanoes and fold mountains. [2 marks]

(c) Earthquakes occur at plate boundaries, where the ocean plate is being subducted. The compression causes faulting and fracturing, which releases energy. They also occur at other places where there is compressive and tensional stress. [1 mark]

Total: 4 out of 6

Examiner feedback

In part (c) one mark has been awarded for *plate boundaries*, but the rest of the answer is explanation, which was not asked for in the question.

1 Volcanoes

Key ideas

- Volcanoes that have erupted in the last 80 years are active.
- Volcanoes that are resting, but may erupt in the future are dormant.
- Volcanoes that are dead and will not erupt again are extinct.
- Active volcanoes occur at certain types of plate margin (see Chapter 10) and in the middle of plates, e.g. Hawaii.
- Volcanoes provide both hazards and opportunities for people.

What comes out of a volcano?

Gases	Liquids	Solids
You should be able to name some of the gases and explain how they control eruptions.	**Lava**	This can be fine dust, such as **ash**, or large rocks.

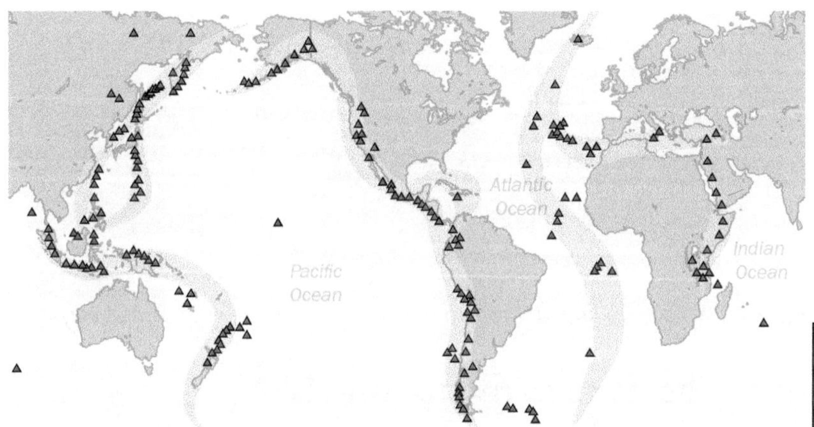

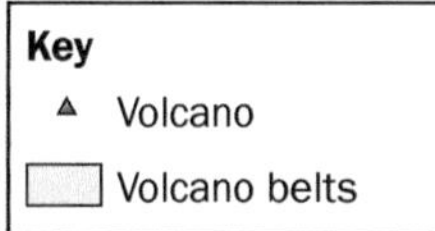

Fig. 11.1 *The world distribution of active volcanoes is not the same as the distribution of earthquakes (see question 1, page 56). Can you explain any differences?*

Types of volcano

Feature	Shield volcano	Stratovolcano
Slope angles	Gentle	Steep
Plate tectonic setting	Constructive margins Mid-plate volcanoes	Destructive margins with an oceanic plate
Products	Mostly lava	Lava and **pyroclastics** (ash)
Lava viscosity	Non-viscous	Viscous
Type of eruption	Continuous and non-violent	Explosive with dormant phases

Exam tip

Make sure you know the features of these two types of volcano and can recognize them from diagrams.

Volcanoes and plate margins

Type of plate margin	Volcanoes	No volcanoes
Collision of continental and oceanic plates	✓	
Collision of two oceanic plates	✓	
Collision of two continental plates		✓
Constructive margins	✓	
Conservative margins		✓

Common error

Remember that earthquakes occur at every type of plate margin, but volcanoes do not.

You should know how melting of rocks in the Earth's mantle occurs and how **magma** rises to produce volcanoes.

Exam tip

Examination questions will often ask you not just to list the hazard, but to describe it and to explain the effect on people's lives, such as their homes or their jobs.

Fig. 11.2 *Volcanoes pose a threat to their local communities in several ways*

Exam tip

You should be able to describe a case study of a volcano using the following headings:

- How the volcano has formed
- Description of an eruption or eruptions
- The hazards to people
- The disadvantages to people

What can be done to reduce the risk?

- Lava flow diversion
- Mudflow barriers
- Building design
- Volcano monitoring
- **Remote sensing**

Hazard mapping and planning

You should be able to describe these features.

Advantages brought by volcanoes

- **Geothermal power**
- Fertile soils
- Volcanoes creating landmasses
- Tourism
- Minerals and mining

For the case study in this part of the syllabus, you should be able to describe an example of a volcano. This should include:

- its origin - how it was formed, including links to plate tectonics
- the hazards or damage caused by the volcano
- any advantages or opportunities that the volcano brought.

You should be able to give specific information about your volcano.

Draw a cross-section of a shield volcano. Label your diagram and list five facts about a shield volcano.

See **www.oxfordsecondary.com/esg-for-caie-igcse** for the answers to the 'Apply' task.

Volcanoes form:

- at **constructive plate margins**, like the Mid-Atlantic Ridge or Iceland
- at **destructive plate margins**, like the Andes and Japan (where subduction occurs)
- occasionally away from plate margins at 'hot spots' like in Hawaii.

Magma is produced deep within the Earth in areas that are hotter than the melting point of the rocks. The magma rises because it is less dense than the surrounding solid rocks.

In more runny magmas (shield volcanoes) the gas is able to escape but in thick, viscous lavas (strato-volcanoes), the gas is released explosively at the surface, producing very violent eruptions which spray lava high into the air. The material then cools and solidifies and falls to the ground. This is how the pyroclastic material (solid) is produced.

Sample question

1. Study the cross-section of a volcano.

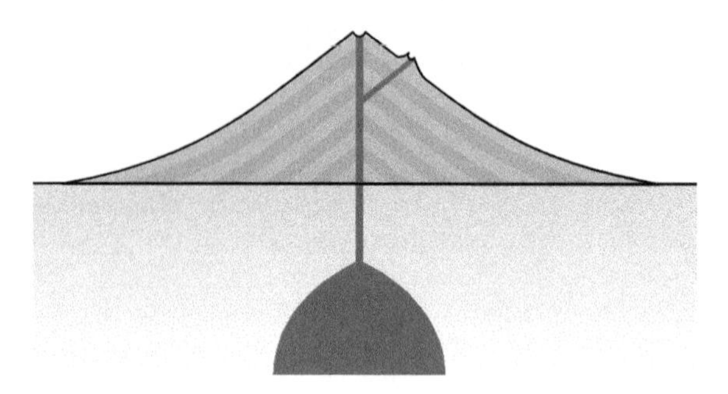

(a) Name the type of volcano shown in the diagram. **[1 mark]**

(b) At what type of plate margins do these volcanoes form? Give an example of this type of plate margin. **[2 marks]**

(c) Describe the shape of the volcano in the diagram and how this is linked to the type of lava. **[3 marks]**

Analysis

✓ You should know the two types of volcano in the syllabus and how they are linked to plate tectonics.

✓ This question is partly answered using the diagram and partly using your own knowledge of volcanoes.

✓ In parts (b) and (c) there are two things to do, not just one.

Student answer

See www.oxfordsecondary.com/esg-for-caie-igcse for the mark scheme for this question.

(a) Stratovolcano [1 mark]

(b) Convergent margins, for example the Himalayas. [1 mark]

(c) This volcano is conical, with steep, concave sides. [2 marks]

Total: 4 out of 6

Examiner feedback

In part (b), the type of plate margin is correct but the Himalayas is not a correct example as there are no active volcanoes there. In part (c), two marks have been scored for the description, but the shape has not been linked to the type of lava.

Key ideas

- Water flows through a drainage basin by different processes.
- A flood or storm hydrograph is a graph showing how a river responds to a rain storm.
- Rivers present opportunities and hazards to people.
- People attempt to manage the effects of rivers.
- There are four processes of river erosion and four processes of river deposition.

Features of a drainage basin

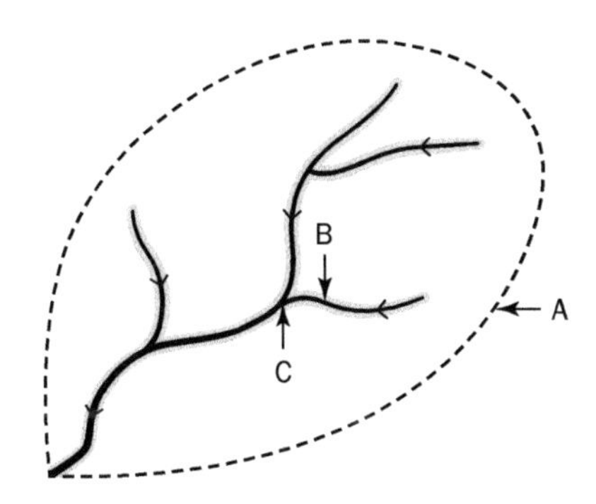

Fig. 12.1 *A drainage basin – the area drained by a river and its tributaries. Identify features A, B and C (you will find them in the glossary at the back of this book).*

Common error

Remember that water leaves a drainage basin by flowing down rivers and by evaporation.

Exam tip

When it rains, it takes a while for water to reach the river. You should be able to explain this process.

Causes of flooding

- Heavy, continuous rainfall on already saturated ground.
- Steep slopes that increase the rate of runoff.
- Impermeable bedrock that rain cannot soak into.
- Urbanization, which creates impermeable surfaces.
- Deforestation, which increases runoff.

Exam tip

You should know about the physical features of rivers (including landforms described in Chapter 13) and the links between rivers and human activity.

Opportunities presented by river valleys and deltas

- Flat land which makes it easy to build roads and settlements and carry out agriculture.
- Soils that are often mineral-rich and fertile, due to the silt and mud deposited by the river during floods (alluvium) – so agriculture is profitable.
- River valleys that are often natural route ways.
- Rivers may be navigable (allowing transport and trade), plus they provide water for drinking and other uses such as irrigation, as well as fish for food.
- In some cases, rivers are used to produce hydroelectricity.

Common error

It is very common for examination candidates to think that rivers flow inland from the sea and split into tributaries. This is completely wrong.

Hydrographs

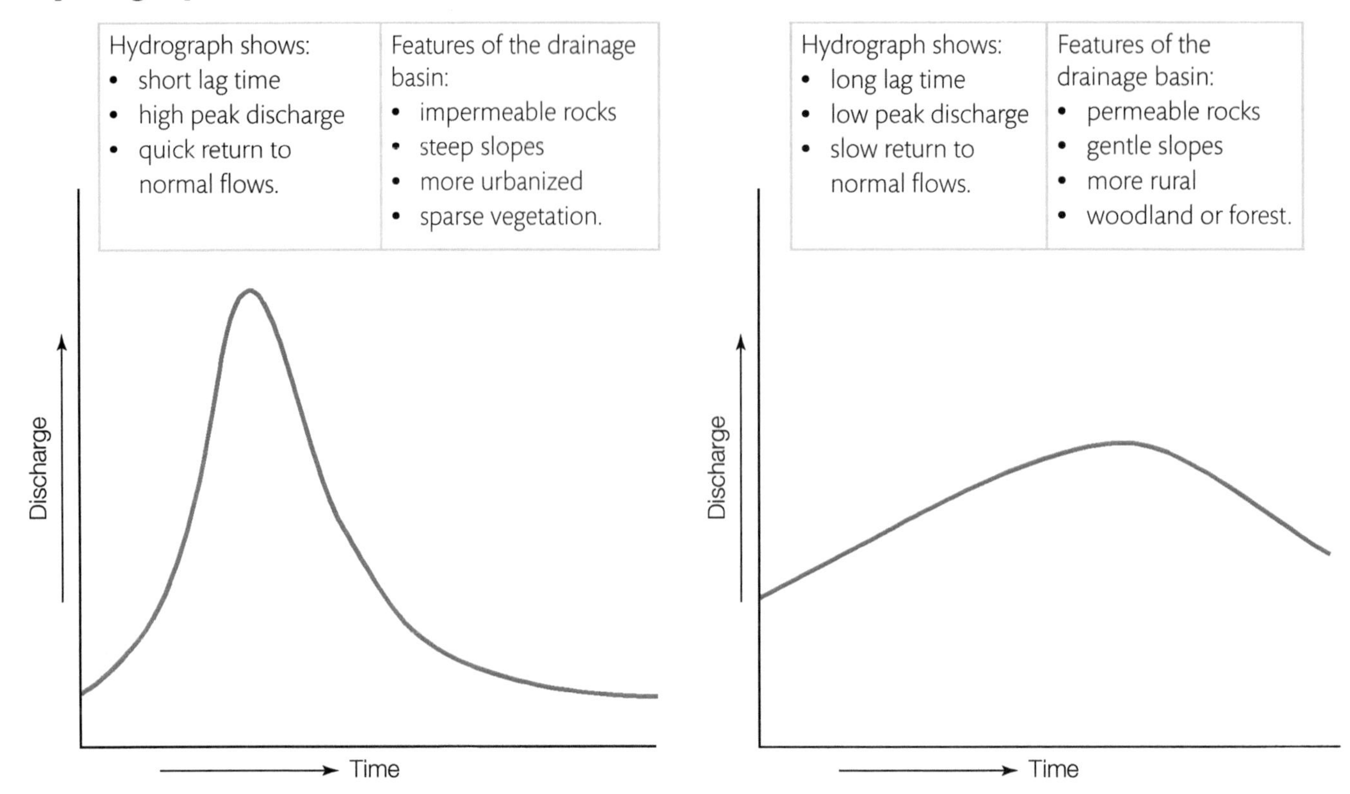

▲ **Fig. 12.2** *The hydrographs of two rivers, affected by the same rainstorm*

Exam tip

Make sure that you can *describe* the shape of a *hydrograph* using the key terms **discharge**, **cumecs**, **rising limb**, **falling limb**, **peak discharge**, **lag time**, and that you can suggest *reasons* for the shape.

Exam tip

Make sure that you can explain why floodplains and deltas are often some of the most densely populated areas on the planet.

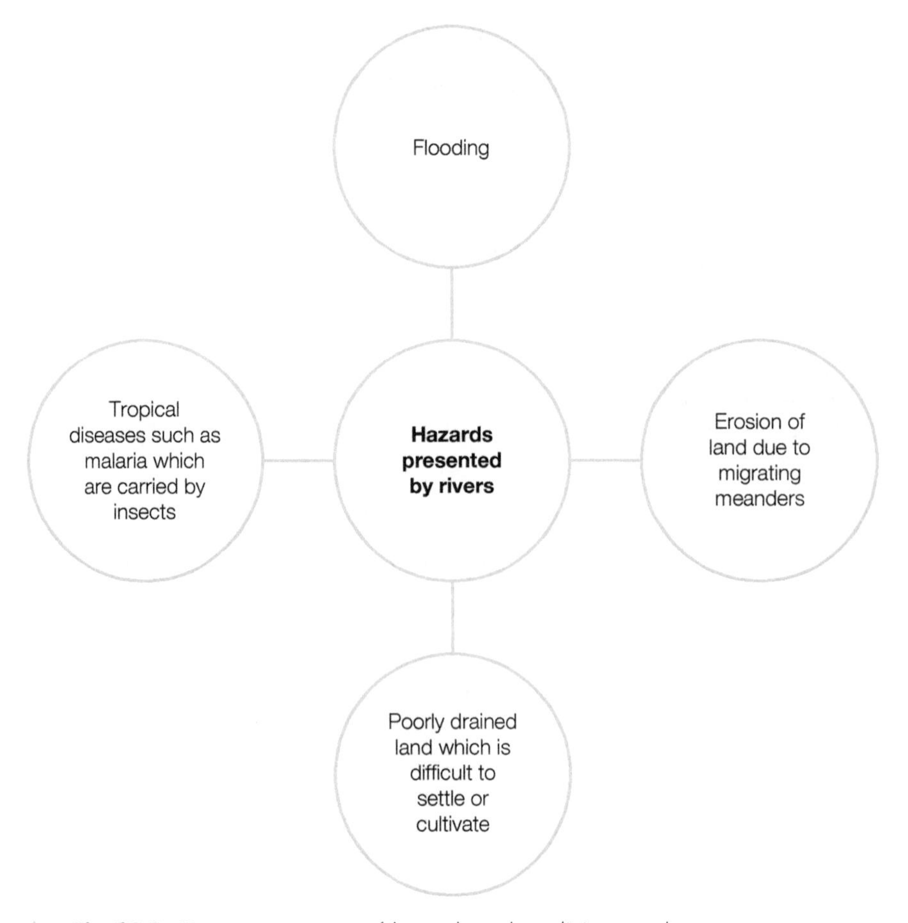

▲ **Fig. 12.3** *Rivers present several hazards to those living nearby*

Managing river flooding and its impacts

You should be able to explain each of the following:

- planting vegetation
- reservoirs
- straightening the channel
- dredging the channel
- artificial levees
- bridge design
- wash lands and building planning.

Exam tip

Make sure that you have a case study, or studies, to illustrate river flooding and the opportunities and hazards of living in river valleys.

River channel patterns

River channels can be **straight**, be **meandering** or be **braided**.

Erosion

There are four processes of erosion: **attrition**, **corrasion**, **corrosion**, **hydraulic action**.

Do not confuse **corrasion** with **corrosion**.

Causes of river erosion

- Increased discharge and velocity
- All the factors causing flooding listed
- Soft rocks
- Channel size and shape; erosion may be concentrated on the outside of meanders, where velocity is greater.

Reducing river erosion

- Re-enforcing river banks
- All the methods of reducing flooding listed.

Transport

There are four processes of transporting a river's **load: saltation**, **solution**, **suspension**, **traction**.

Common error

Do not confuse the four processes of *transportation* with the four processes of *erosion*.

Deposition

Deposition occurs when a river loses velocity (energy). This can be caused by:

- a decrease in gradient
- a decrease in river flow (discharge) as water drains away after heavy rain
- the river meeting the sea or a lake, often forming a delta
- the river flowing more slowly on the inside of bends.

- Water flows through a drainage basin by **interception, infiltration, throughflow** and **groundwater flow**.
- Flood or storm hydrographs have different shapes depending on rock permeability, slope angles, vegetation type and amount of urbanization.
- People live in river valleys because of the opportunities they present, but face hazards of flooding and erosion.
- People attempt to manage the hazards caused by rivers.
- Remember that there are four processes of river erosion and four processes of river deposition.

Name the odd one out in each of the following lists:

(a) corrosion, corrasion, attrition, saltation
(b) hydraulic action, traction, suspension, solution
(c) interception, infiltration, groundwater flow, watershed
(d) energy, velocity, discharge, load
(e) meandering, V-shaped, braided, straight.

Water flow processes	Erosion processes	Transport processes	Channel shapes
interception	hydraulic action	traction	meandering
infiltration	corrasion	saltation	braided
throughflow	corrosion	suspension	straight
groundwater flow	attrition	solution	

Raise your grade

Sample question

1. For a major river valley that you have studied, explain the opportunities and disadvantages of living there. **[7 marks]**

Analysis

✓ It is important that your answer should cover both the opportunities and the disadvantages.

Student answer

People live in the Indus Valley in Pakistan for various reasons. The river and its tributaries are the main water supply for 170 million people. Rainfall in southern Pakistan is less than 250 mm per year, but there are rich alluvial soils. There are irrigation dams and canals, including the Guddu Barrage. Pakistan produces large amounts of wheat and cotton. Hydroelectric projects, e.g. the Taunsa Barrage are very important for industrial and domestic supply.

Industrial pollution is affecting the vegetation and wildlife of the delta. The river may shift its course destroying land, although the change may take centuries. Sediment is clogging irrigation canals, affecting agricultural production. The irrigation water evaporates, leaving salt deposits that make land infertile. Floods can have a devastating impact. In 2010, 1700 people drowned and 14 million people had to leave their homes.

[Level 3: 7 out of 7]

Examiner feedback

This is an example of a short answer which would score high marks. Almost all the statements made are fully developed. The example is correct and there is a lot of specific detail about the Indus Valley.

See www.oxfordsecondary.com/esg-for-caie-igcse for the mark scheme for this question.

Sample question

2. Choose an example of a river valley and describe the causes of flooding in the area you have chosen. **[7 marks]**

Analysis

✓ Your answer should include at least three causes and should explain each in detail.

✓ Choose an example where you can give local details.

Student answer

In the Indus Valley in Pakistan major flooding has been caused by very heavy rainfall and melting snow and glaciers. Levees have broken causing flooding. Deforestation has also been a cause.

[Level 1: 3 out of 7]

Examiner feedback

The answer contains four simple statements but none of them are developed so this is a Level 1 answer.

All the points mentioned in the answer are relevant, but each point is not properly explained and there is no information which is specifically about Pakistan. Reference could be made to the summer monsoon rains or by quoting figures, e.g. Hyderabad recorded 77 mm of rain in 24 hours on 7 August 2016. Melting snow and deforestation are in the Himalayas. In some countries flood risk has been reduced by building levees along rivers. Levees and deforestation have caused the river channel to silt up, causing even bigger floods.

See www.oxfordsecondary.com/esg-for-caie-igcse for the mark scheme for this question.

Key ideas

- In the upper course of a river, the gradient is steep and the river is carrying out vertical erosion.
- In the middle and lower course, the gradient is more gentle and the river is carrying out lateral erosion and deposition.
- Landforms of the upper course include potholes, rapids, waterfalls, gorges and interlocking spurs.
- Landforms of the lower course include meanders, oxbow lakes, floodplains, levees and deltas.

Common error

Although the gradient in the upper course may be steep, it doesn't mean that the water is flowing any faster. In fact, friction from the bed and banks may mean that the water is flowing more slowly.

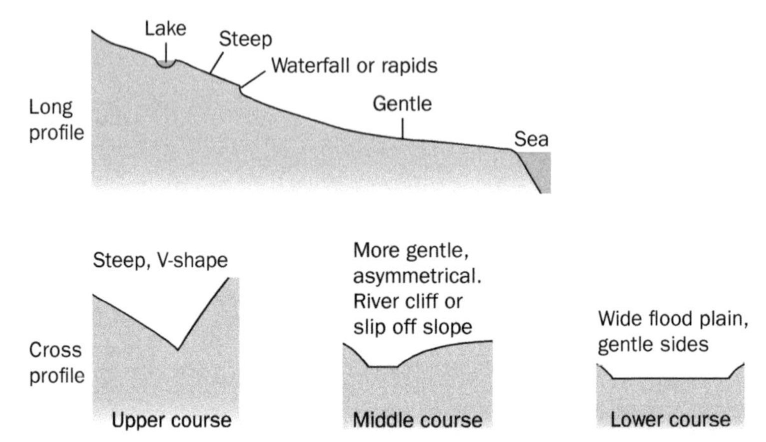

Fig. 13.1 *The long profile of a river and the cross profile of a valley*

Exam tip

For the landforms listed here, you should be able to *describe* them and *explain how they were formed.*

Landforms of the upper course of a river

The place where a river starts its course is known as the source. The shape of the valley in cross-section is a V-shape. The valley floor is very narrow and the river channel may occupy the whole of it. Large boulders in the river – the bed load – are only moved after heavy rainfall when the river becomes a powerful torrent.

Landform	Description	How it is formed
Potholes	Smooth, rounded hollows in the bedrock of the bed of the channel. They are often about 30 cm across.	They are formed by stones being trapped in hollows on the bed of the river. Eddies in the water swirl the stones around causing corrasion. The hollows become deeper and wider and eventually join together.
Rapids	Places where the water is shallow and the bed of the river is rocky and irregular. The gradient is often steeper than at other points in the river's course. Rapids are often a barrier to navigation on rivers.	Rapids may be caused by a band, or bands, of hard rock in the riverbed.
Waterfalls and gorges	A gorge is a deep, steep-sided valley with a waterfall.	Formed when a horizontal layer of hard rock lies over a layer of soft rock. The soft rock is eroded more quickly. Gradually a **plunge pool** develops. The splashing water and eddy currents in the plunge pool undermine the hard rock. The hard rock layer is left unsupported and eventually collapses. If the processes of undercutting and collapse are repeated over a long period of time, the waterfall will retreat upstream forming a gorge.
Interlocking spurs	In the upper course of a river, where the valley is narrow, there are spurs of land on either side of the valley and the river winds around them.	

Landforms of the middle and lower courses of a river

The valley is wider, with more gentle sides and a floodplain.

Landform	Description	How it is formed
Meanders and oxbow lakes	Meanders are severe bends in a river and oxbow lakes are sections of isolated river, close to the original river, which have been cut off.	They are formed by erosion on outer bends where water flows fastest, which forms a river cliff. Deposition on inside bends, where water flows slowest forms a slip-off slope. This increases the size of meanders until a narrow neck forms. The river breaks through the neck during flooding. The ends of an oxbow lake are sealed by deposition.
Floodplains	The land next to a river which floods is called a floodplain.	Their formation begins with deposition of point bars on the insides of meanders. These deposits are spread across the valley as the meanders migrate. There is deposition of gravel on the riverbed (part of the bed load). There is deposition of fine silt and mud (part of the suspension load) on the floodplain itself during floods when the river overflows its banks.
Levees	Levees are natural ridges of sediment on the banks of a river.	Levees are formed when the river floods. As the water overflows the channel, it slows down (loses energy) and deposits more of the load (and a coarser part of the load) close to the channel, making the banks higher. During normal flows, the river deposits on the riverbed within its channel, making the bed higher than the surrounding floodplain.
Deltas	The delta of a river is its mouth, where it meets the sea.	Deltas form because the river carrying its load meets the still water of the sea or lake. The loss of velocity (energy) leads to deposition which builds up to form the delta. Over time, the delta is gradually built out into the sea. Deposition blocks the channels, which leads to the formation of distributaries.

Exam tip

You should be able to illustrate your descriptions of the landforms with *labelled* diagrams.

Common error

Some examination questions ask about features of the *river*, while others ask about features of the *valley*. They are not the same thing.

Exam tip

When explaining how these landforms were formed, remember to include the processes of erosion and deposition described in Chapter 12.

Common error

Candidates often think that the floodplain is only formed from deposition from floods. This is not true. There are also deposits on the inside bends of meanders, and gravels deposited on the bed of the river channel.

Recap

- Landforms of the upper course of a river include potholes, rapids, waterfalls, gorges and interlocking spurs. These are formed by erosion, which is mainly vertical.
- In the upper course of a river, the gradient is steep and the long profile is irregular.
- Landforms of the lower course of a river include meanders and oxbow lakes, floodplains, levees and deltas. These are formed by lateral erosion and deposition.
- In the middle and lower course, the gradient is more gentle and the long profile is smooth.

See www.oxfordsecondary.com/esg-for-caie-igcse for the answers to the 'Apply' task.

Apply

Name the landform at A and the landform at B.

Name the process at A and the process at B.

Which course of the river is shown?

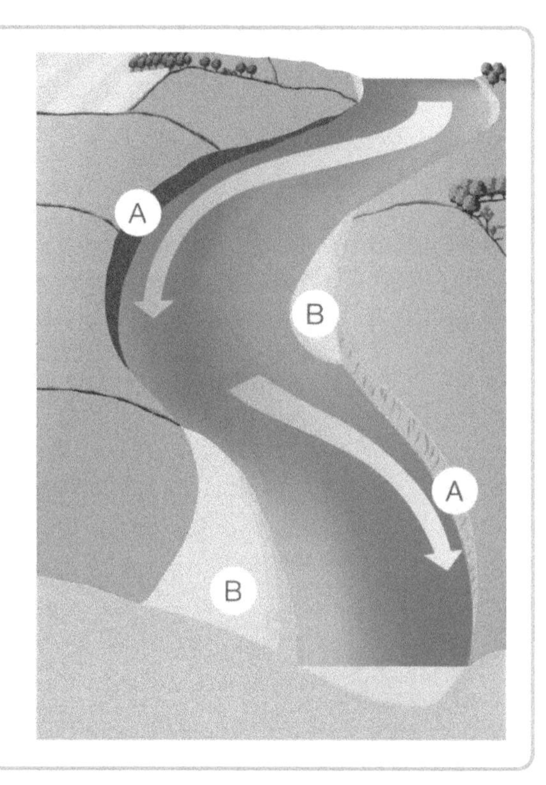

Review

	Upper course	Middle and lower courses
Gradient and long profile	Steep and irregular with lakes, waterfalls and rapids	Gentle and smooth
Cross profile	Steep and V-shaped	Gentle sides, flood plain develops
Processes	Vertical erosion	Lateral erosion and deposition
Landforms	Potholes, rapids, waterfalls, gorges and interlocking spurs	Meanders and oxbow lakes, floodplains, levees and deltas

Sample question

1. Study the diagram which shows two river valleys.

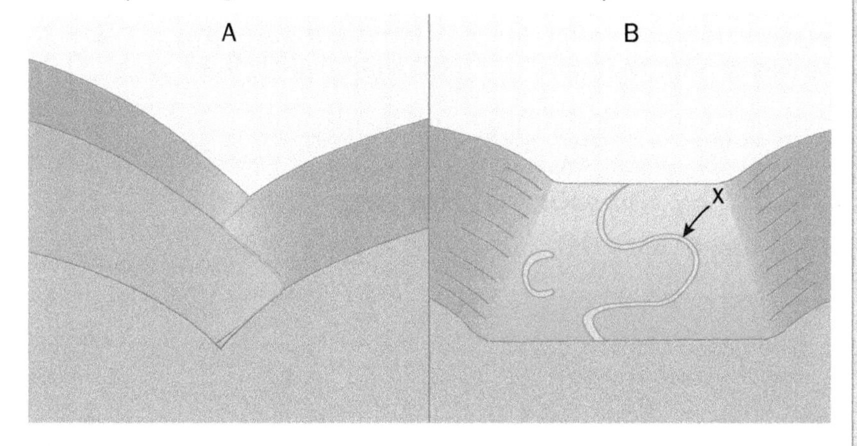

(a) Name the feature of the river at X on the diagram. **[1 mark]**

(b) Which part of the river valley does each diagram show, the *upper course*, the *middle course* or the *lower course*? **[2 marks]**

(c) Using the diagram, name **one** feature of river valley A and **two** features of river valley B. **[3 marks]**

Analysis

✓ This question tests your knowledge of river landforms. The following landforms are named in the syllabus: waterfalls, potholes, meanders, oxbow lakes, deltas, levees and floodplains. You should be able to identify each one in a diagram or a photograph.

Student answer

(a) Meander [1 mark]

(b) A is the upper course and B is the lower course. [2 marks]

(c) A interlocking spurs and B floodplain and oxbow lake. [3 marks]

Total: 6 out of 6

See www.oxfordsecondary.com/esg-for-caie-igcse for the mark scheme for this question.

Examiner feedback

This is a completely correct answer.

Key ideas

- Coasts are shaped by the work of the sea and wind.
- The sea forms landforms by eroding, transporting and depositing.
- Destructive waves erode and constructive waves deposit sediment.
- Sediment is transported along the shore when waves reach it obliquely.
- Erosion occurs on exposed coasts, such as on headlands.
- Deposition takes place in calm, sheltered areas like bays and inlets.

Erosion by the work of the sea (marine erosion)

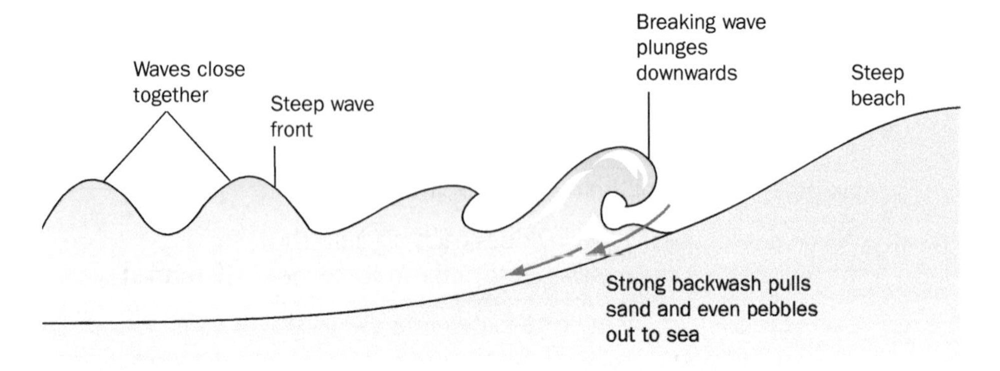

Fig. 14.1 *Destructive waves steepen beach profiles*

Common error

Remember that caves, arches and stacks form at lines of weakness in rocks, not in layers of soft rock.

Landforms formed by erosion by the sea

Landform	Description
Cliffs	Cliffs are vertical or very steeply sloping rocks on the coast. Cliffs in horizontal rocks are vertical. If the rock layers dip steeply away from the sea the cliff profile slopes more gently. Many cliffs have an indentation called a wave-cut notch at about the high-tide level where wave attack has undercut them. The rock above the undercut eventually collapses and the cliff wears back, leaving a wave-cut platform.
Wave-cut platforms	These solid rock platforms between the cliff base at high-water mark and the low-water mark slope gently towards the sea and may have rock pools. They are often covered with debris eroded from the cliffs.
Headlands	Headlands are areas of resistant rock projecting out to sea. They have cliffs along their sides.
Bays	Bays are approximately semi-circular shaped areas of sea. The land behind them is lower and more gently sloping than the headlands to either side of them.
Caves	These are indentations at the base of cliffs, formed where lines of weakness have been enlarged by erosion.
Arches	Arches are holes right through headlands and are formed by the joining of caves eroded into both sides of the headlands, along a line of weakness.
Stacks	These vertical pillars of rock are isolated in the sea off headlands (to which they used to be attached). They have flat tops if the rocks are horizontal. Others have pointed tops.
Stumps	The worn down low remnants of stacks uncovered at low tide.

Transportation by waves: Longshore drift

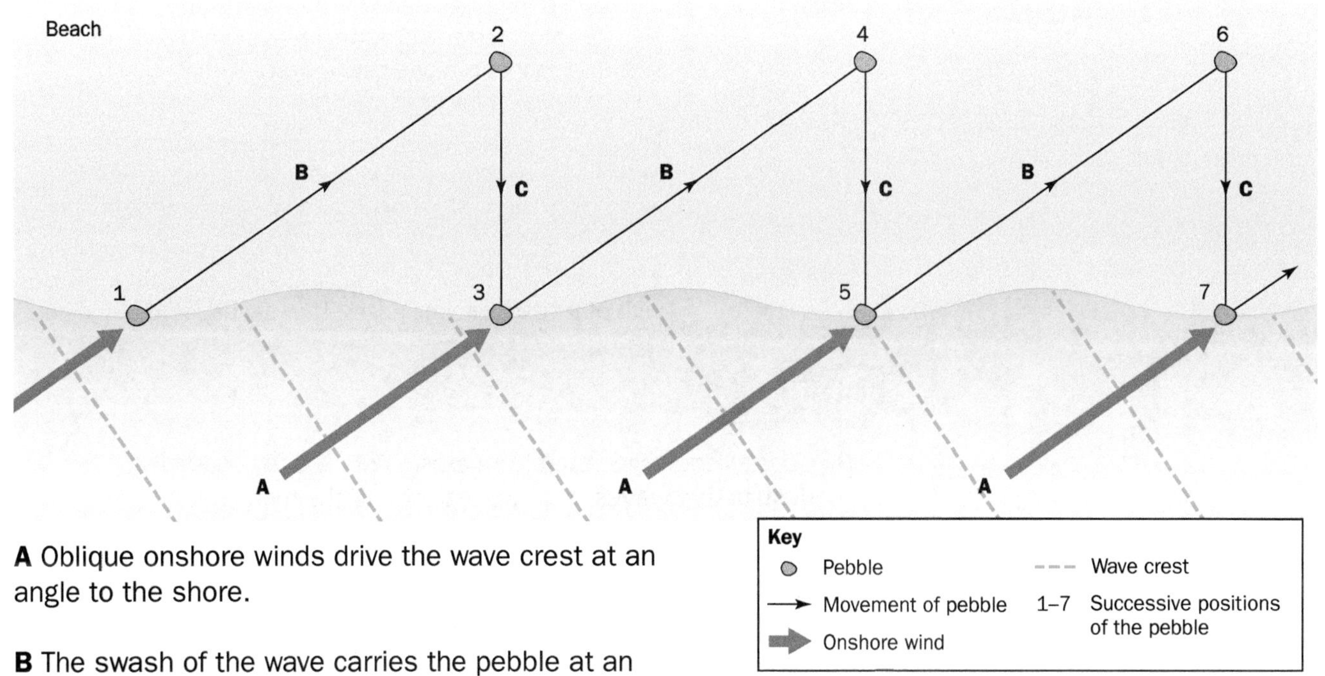

A Oblique onshore winds drive the wave crest at an angle to the shore.

B The swash of the wave carries the pebble at an oblique angle up the beach.

C The backwash of the wave brings the pebble straight down the beach under the influence of gravity. As this process is repeated, the pebble is moved along the beach. The direction of longshore drift is from left to right in this example.

Fig. 14.2 *Movement of sediment along the beach*

Landforms formed by deposition by the sea (marine deposition)

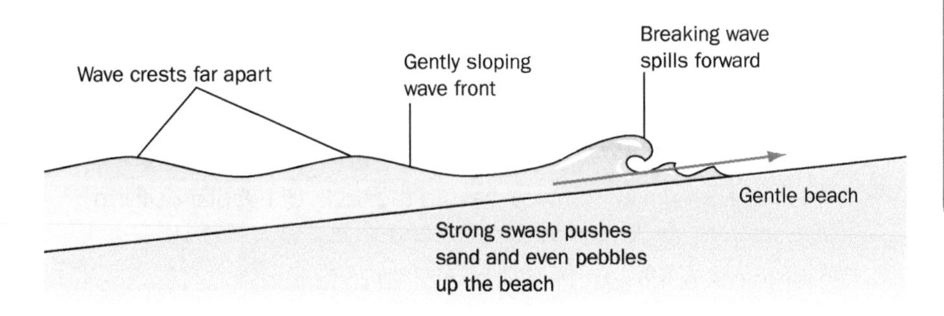

Fig. 14.3 *Low, gentle constructive waves deposit sediment to form beaches and spits of sand and shingle*

Waves sort material into different sizes. Often beaches have coarser shingle at the top, sand lower down and fine mud beneath low-tide level. They may have sand dunes above them formed of sand removed from the beach and deposited by wind.

Common error

There are no landforms caused by longshore drift, but it is a very important process bringing sediment for a landform to develop by deposition.

Exam tip

When describing landforms remember to think of all aspects. What is it made of? What shape is it in plan? What shape is it in profile? How large is it in area or height?

Landform	Description
Beach	Sand and shingle between low- and high-water marks. Shingle forms a steep slope, whereas sand beaches are gently sloping. Beaches can be straight, crescent-shaped in bays, and triangular at the head of inlets.
Spit	Long, narrow, low ridge of sand or shingle deposited at a bend in the coast, such as a bay or river mouth. Spits are attached to the land at one end and the other ends in open water. They may have re-curved or hooked ends.
Sand dune	A sand hill or ridge of sand parallel to the shore, usually found with other parallel dune ridges. They are deposited by wind.

Recap

- Coastal landforms result from erosion of exposed coasts by **destructive waves** and by deposition of the material by **constructive waves** in sheltered waters.
- During transport the load is eroded down and rounded by attrition to shingle and then sand.
- Headlands and bays form by differential erosion of rocks that lie at right angles to the coast.

See www.oxfordsecondary.com/esg-for-caie-igcse for the answers to the 'Apply' task.

Apply

1. Draw a simple diagram of a headland which has a wave-cut platform, stump, stack, cave and arch. Label each landform to describe its features.
2. Explain how the landforms cave, arch, stack and stump are formed in a sequence.

Review

Some coastal landforms are formed in a sequence:

- Waves undercut coastal rocks to form cliffs. Cliffs retreat as they are undercut to leave a wave-cut platform.
- Waves attacking a line of weakness in the rocks of headlands form caves, arches, stacks and stumps in sequence.
- Beaches can extend at bends in the coast to form spits as longshore drift supplies sediment for deposition.
- Onshore winds move beach sand to the top of the beach to build sand dunes.

As the characteristics and formation of landforms are easier to understand and remember if they are seen, diagrams and photographs are very helpful. There is a comprehensive coverage of these in Chapter 5 of *Complete Geography for Cambridge IGCSE and O Level*.

Sample question

1. Study the photograph of a coast.

(a) Name and describe a landform of coastal deposition shown in the photograph. **[1 mark]**

(b) Identify **two** landforms of coastal erosion shown in the photograph. **[2 marks]**

(c) Explain why the sea erodes some areas of rock back more than others. **[3 marks]**

Analysis

✓ This question requires you to understand how the degree of coastal erosion depends on the interaction of waves and the characteristics of the rock.

✓ When asked to name a feature shown in a photograph, give the name of the feature that is given in the syllabus, not a local name.

✓ The photograph shows the coast at low tide, so it will help you to picture it at high tide when the sand is covered by water to help you recognize the stump and bay.

✓ Photographs are not usually studied sufficiently closely. Describe what you actually see, not what you think there should be in the photograph.

Student answer

(a) Beach with destructive waves breaking on it [0 marks]

(b) It is a coast with cliffs and has a cave. [1 mark]

(c) The less resistant rocks are more easily eroded than the resistant rocks, so they are worn back to form bays. The harder rocks resist erosion, forming headlands which protrude out into the sea between the bays. [3 marks]

Total: 4 out of 6

Examiner feedback

For part (a), the waves are not part of the beach, so the required description is missing. In part (b), the cliffs are correct for one mark, but it is incorrectly assumed that a cave can be seen on the photograph. In part (c), three good points are made.

See www.oxfordsecondary.com/esg-for-caie-igcse for the mark scheme for this question.

15 Coasts (2)

Key ideas

- There are three types of coral reef: fringing reef, barrier reef and atoll.
- Coral reefs lie parallel to the coast. They have breaks in them, usually at the mouths of rivers.
- Coral can only live in warm, specialized conditions, usually within 30° of the Equator and within 30 metres of the sea surface.
- Coral reefs are useful and valuable to people.
- They are very fragile, easily destroyed ecosystems.
- Mangrove swamps are only found between 32°N and 38°S.
- Mangrove swamps are very valuable to people in many ways.
- Coasts offer opportunities for economic development and leisure.
- Coasts are a hazard for people because they erode and are badly affected by storms.

Coral reefs

Diagram	Reef type	Description
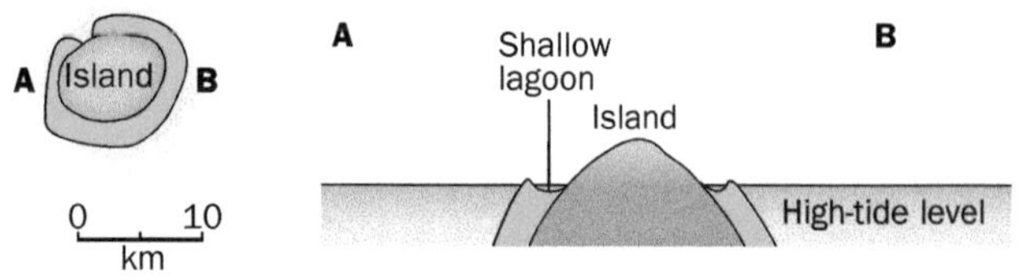 A Island B; 0 10 km; A Shallow lagoon Island B High-tide level	**Fringing reefs, e.g. on the Coral Coast of Fiji**	A low, narrow band of coral next to, and parallel to, the coast at about low-tide level, which is covered by a narrow, shallow **lagoon** at high tide. Beyond it, the higher outer edges of the reef rise to about high-tide level and its outer edge slopes steeply down into the sea.
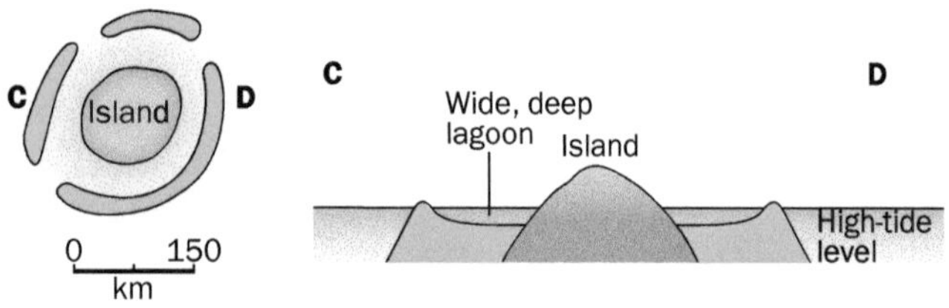 C Island D; 0 150 km; C Wide, deep lagoon Island D High-tide level	**Barrier reefs, e.g. the Great Barrier Reef of Australia**	Usually several kilometres from the coast (a minimum of 500 metres), separated from it by a wide, deep lagoon. The Great Barrier Reef consists of almost 3000 reefs, separated by channels, stretching more than 2600 km.
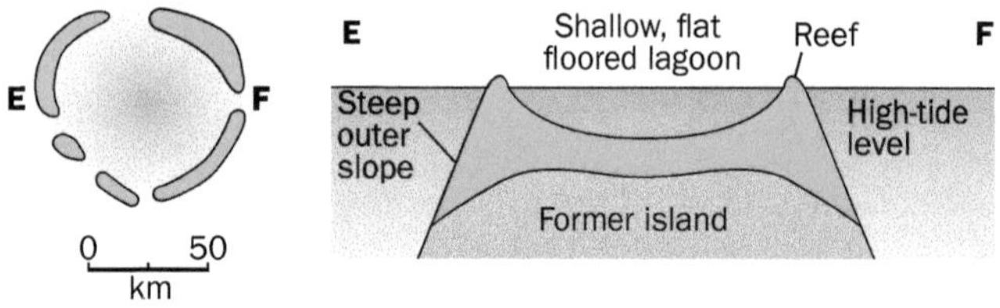 E F; 0 50 km; E Shallow, flat floored lagoon Reef F Steep outer slope High-tide level Former island	**Atolls, e.g. Suvadiva Atoll in the Maldives**	Narrow, circular reefs, broken by channels. They surround a deep lagoon. The lagoon can be more than 50 km wide. The Maldives are a north-to-south chain of atolls, 13 of which are very large.

Conditions required for the development of coral reefs

- Warm water – reefs grow best where the mean surface water temperature is 22–25 °C and cannot live in sea temperatures lower than 18 °C.
- Oxygen and food – the outer edge of the reef is highest where oxygen and food are most abundant due to the waves that break there. Polyps cannot survive long periods above the water so much of the reef is at low-tide level.
- Clean, clear unpolluted water – they cannot live where rivers deposit sediments or pollutants into the sea. Sediment cover prevents them feeding.
- Conditions of normal salinity.

- Abundant sunlight – needed by corals and the single-celled algae that live with them and which the **coral polyps** cannot live without, so shallow waters of about 10 metres depth are best.
- A solid surface – such as a submerged island, from which the reef growth started.

Coral reefs are damaged by polluted waters and the Crown of Thorns starfish which thrive in them; warmer seas; storms; contact with boats and people.

Mangrove swamps

Conditions required for the development of mangrove swamps

- Warm conditions where the sea surface is no lower than 16 °C and air temperatures are at least 20 °C.
- Tidal areas, between low-tide and high-tide marks.
- No strong waves or tidal currents.

Ways in which mangrove are adapted to live in their environment

Adaptation	Reason for the adaptation
Roots that filter out salt and leaves that excrete salt	Enables them to live in brackish water
Conical breathing roots (aerial roots) stick vertically up out of the water at low tide	Enables them to live in submerged soils without oxygen
Lateral spreading roots which rapidly extend out from seedlings	Anchors them in the soft mud
Prop or stilt roots extend down from the trunk	Gives them stability and strength as tides rise and fall, and when strong waves attack or storm surges occur

Value of mangrove swamps

- **Mangrove swamps** stabilize the coast against erosion during large waves by their firmly anchored roots.
- They protect the coast from flooding during storms and tsunamis.
- They give some protection from the strong winds of tropical storms, such as hurricanes.
- They absorb inorganic nutrients that drain into them in water from farmland and urban areas, preventing the nutrients from harming marine life.
- Mangrove swamps catch sediment washed towards the sea after heavy rains, preventing it from harming coral reefs.
- They are important habitats for many species, including crocodiles, frogs, snakes and fish. They are also important habitats and breeding grounds for birds, such as parrots, heron and frigate birds. Small fish feed on organic nutrients in the water produced by the decay of mangrove leaves.
- They form a safe nursery for young fish, shellfish and baby turtles, which can hide among the roots.
- They are valuable to local people and tourists for recreation: fishing, bird watching, wildlife photography and boating.

Coasts present hazards and offer opportunities to people

Opportunities that coasts present for people

- Recreation and well-being, for example water sports, beach activities, walks, swimming, diving.
- Activities which are good for the economy, such as tourism, sites for ports to stimulate trade and industry, fishing and the extraction of salt and rock materials.
- Coastal lowlands offer good sites for settlements and route-ways.

Common error

For a question about human activity, remember that 'coast' can include a wide strip of land next to the coastline, e.g. a lowland that could be flooded by storm waves. Small coastal countries could be acceptable for a case study, as could the southern half or delta of Bangladesh.

Hazards that coasts present for people

Erosion

Destructive wave action at the foot of cliffs can undermine buildings and roads, causing them to collapse into the sea, causing financial losses to people and authorities.

Measures that may be taken to reduce erosion if the benefits outweigh the costs	
Soft engineering (helping the natural protective processes)	**Hard engineering** (using structures to reduce the power of the wave)
Beach nourishment which keeps the beach as a buffer between cliffs and the sea	**Groynes** which encourage beach material to accumulate in front of the cliffs
Conservation of sand dunes to keep a buffer between the land and the sea	Sea walls to reflect the waves
Pipe drainage through cliffs to remove the weight of water and minimize the danger of collapse	**Gabions** and **rip-rap** to dissipate wave energy; **revetments** to absorb their energy

Tropical storms

Tropical storms cause great destruction of infrastructure and affect livelihoods and the well-being of people living on coastal lowlands because of:

- very strong winds
- flooding caused by excessive rains and **storm surges**.

Difficulties of managing tropical storms

Background information about tropical storms is included in Chapter 6: Weather in the Student textbook, pp.175–179 for example.

- It is difficult to be certain where to evacuate until just before a storm strikes because the storm path can change.
- The violence of the wind and rise of sea level in a storm surge cannot be controlled.
- Strong storm shelters may help save lives if people can be given enough warning to get to them and if there are enough within easy reach of the community.

- Forecasting and evacuation is easier in an MEDC, where there is satellite technology to watch the path of the weather system and government agencies are able to issue warnings to evacuate. In LEDCs, these may not always be so quickly available.
- The expense of providing emergency shelters and stocking them with supplies is more easily met in MEDCs than in LEDCs.
- Management in LEDCs is often limited to rescue, aid and restoration after the storm, often with the help of MEDC aid.
- Unlike in MEDCs, homes and businesses in LEDCs are often not insured against storm losses.

Exam tip

You should be able to describe the opportunities presented by an area or areas of coastline, the associated hazards and their management.

Recap

Coral reefs, mangrove swamps and tropical storms are found only in warm subtropical and tropical latitudes. Tropical storms are largely confined to east coasts.

Coral reefs are built up from the skeletons of coral polyps where the sea water is about 22–25 °C, clean, shallow and clear, with abundant sunlight and oxygen.

Coral polyps are killed by pollution, being covered by sediment or touched by human activity and by warming of the ocean.

Mangrove forests develop on soft mudflats in warm, sheltered, tidal waters. Special adaptations enable mangroves to live in the difficult environment. Their roots filter out salt and leaves excrete it. Other roots stick out of the water at low tide for breathing, and prop or stilt roots give stability.

Both ecosystems are very useful for people and all coasts offer opportunities, but erosion and tropical storms are hazardous.

- Damage caused by coastal erosion and by the impacts of tropical storms and their ferocious winds, torrential rain and storm surges, is very expensive to repair.
- Damage is usually greater in LEDCs than MEDCs, in particular because of greater protection, planning and forecasting in MEDCs, as well as the flimsier structures in LEDCs.

Apply

Describe ways in which coral reefs and mangrove swamps offer similar opportunities for people, and how their value for people varies.

See www.oxfordsecondary.com/esg-for-caie-igcse for the answers to the 'Apply' task.

Review

- Coasts provide job opportunities in fishing, tourism, ports and harbours.
- They boost a country's economy as ports stimulate the growth of industry. Salt, sand, gravel and rocks can be extracted and sold. Tourism generates high income.
- Coasts give opportunities for many leisure activities.
- Protective measures against erosion are usually put into place only if the expense is justified by the value of what they will protect.
- Measures vary from expensive hard-engineering projects that build structures, to soft-engineering projects that help natural processes to work.

There are illustrations of these features in Chapter 5 of *Complete Geography for Cambridge IGCSE and O Level*. Case studies include the coast of the Cayman Islands and of a tropical storm in Fiji.

Fig. 15.1 *Hard engineering on the coast of Turkey (see question 1 opposite)*

Sample question

1. Study Fig. 15.1, an enlargement of this photograph of a coast in Turkey.

(a) State **one** physical feature that shows that erosion has affected the coast in the photograph. **[1 mark]**

(b) Describe the methods being used to try to reduce erosion. **[2 marks]**

(c) Describe **two** other methods of reducing coastal erosion and briefly explain why they are used. **[4 marks]**

Analysis

✓ Locate features on the photograph by using 'left', 'centre' or 'right'; and 'foreground', 'middle ground' and 'background'; or combinations of these.

✓ Think about the evidence on the photograph for the coastal protection methods being used.

✓ Notice when a requirement is in the plural to score full marks.

Student answer

(a) Rocks in the foreground. [0 marks]

(b) The lorry has brought large rocks which are being put at the side of the low cliff by the bulldozer as a buffer against the sea. Rip-rap is being laid down. [1 mark]

(c) Gabions, rocks in wire baskets, could have been put in a line in front of the cliff to absorb the power of the waves. A revetment or sloping concrete wall could have been used there instead. [3 marks]

Total: 4 out of 7

Examiner feedback

The answer to part (a) is much too vague. Which rocks? It needs to be more specific, such as there is a small, isolated island in the left middle ground of the photograph. Part (b) asks for 'methods', and this is all one method, so cannot gain full marks. A new wall has also been built, presumably to keep the sea from flooding behind it. The answer for part (c) is good, but incomplete and would not gain full marks because the use of the revetment has not been explained.

Key ideas

- Weather is the state of the atmosphere at a particular time.
- Weather recordings can be read by observation from an instrument or can be automatically and continuously transmitted to a computer from a digital instrument.
- Weather readings taken manually should be recorded at the same time each day, usually at 9 a.m.
- The eye should be at the same height as the marker that is being read to avoid distortion (with the necessary exception of the wind vane).

Common errors

Remember:

- that the readings of maximum and minimum temperature on the Six's thermometer are read from the bottom end of each index.
- that the arrow showing wind direction points to where the wind *has come from*. That is the wind direction. Express it clearly, e.g. 'from the west' or 'a west wind', not 'it is in a westerly direction', which is ambiguous.
- be careful to spell hygrometer with a 'g' not a 'd', as there is a different instrument with that spelling.

Weather elements and weather station recording instruments

Weather element	Manually read recording instrument	Unit of recording
Precipitation	Rain gauge	Millimetre (mm)
Temperature	Six's thermometer (maximum-minimum thermometers)	Degrees Celsius (°C)
Humidity	Hygrometer or wet- and dry-bulb thermometer	°C (used to calculate the percentage **saturation** of the air using tables)
Pressure	Aneroid barometer for a particular time or barograph for a record of changes over 24 hours	Millibar (mb)
Sunshine hours	Sunshine recorder (Campbell-Stokes recorder)	Hours/minutes
Wind direction	Wind vane or wind sock	Cardinal points from which the wind has come
Wind speed	Anemometer	Kilometres per hour (km/h)

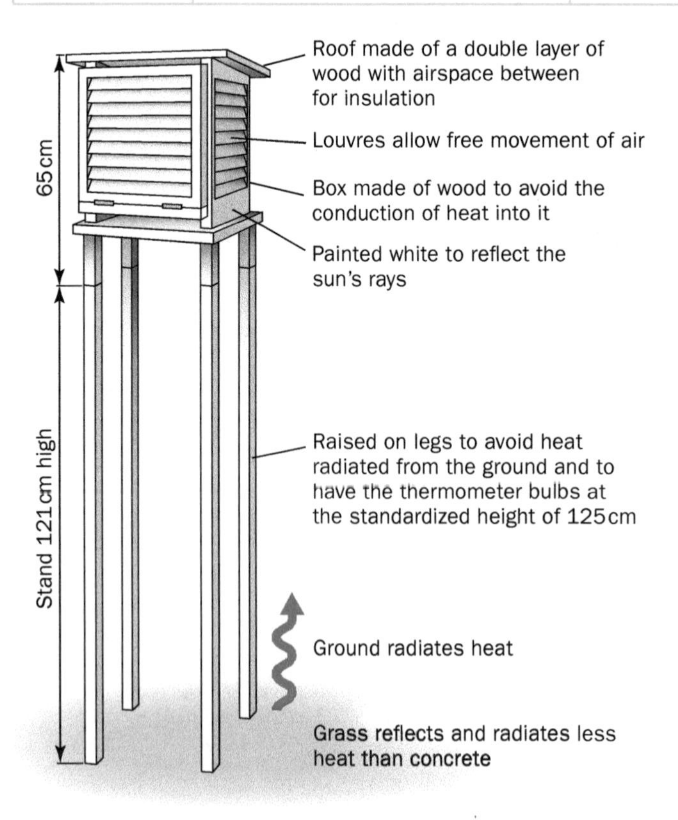

Fig. 16.1 *Thermometers are housed, often with the barometer, in the Stevenson screen, which is designed to ensure that the instruments give the correct readings*

Weather station layout

- Thermometers must be on grass and away from buildings which may radiate heat.
- Wind vanes and anemometers should be in an open space away from any trees or buildings. They must be at least three times the height of the nearest obstacle to avoid being sheltered.
- The rain gauge must be in an open space. It must be at least twice the height of the nearest obstacle to avoid being sheltered and rain drip.
- Barometers and barographs should be kept away from strong air movements, direct sunlight and heat sources (heat causes air expansion, lowering the pressure).
- The sunshine recorder should be in an open space, without any shade and placed to face the sun.

Exam tip

Know how weather data are collected from specific instruments. You should be able to describe the instruments and where they should be sited to give accurate readings and the reasons for the uses and construction of the Stevenson screen.

Digital weather recording instruments

Weather element	Digital recording instrument
Precipitation	Tipping bucket rain gauge that tips when full. The amount is calculated by the known volume of the bucket multiplied by the number of tips transmitted, e.g. every 30 minutes to a computer to give a continuous day's record.
Temperature	Maximum-minimum thermometers with the reading on a screen.
Humidity	Hygrometer with reading in percentage figures on the screen.
Pressure	Has a corrugated metal box which transmits to a computer.
Sunshine hours	A sunshine sensor which logs the duration and intensity of the sunshine and transmits it to a computer.
Wind direction	Digital wind vane which has a paddle to catch the wind. The direction is shown by the stalk.
Wind speed	Anemometer with three rotating cups and the speed is shown on a screen.

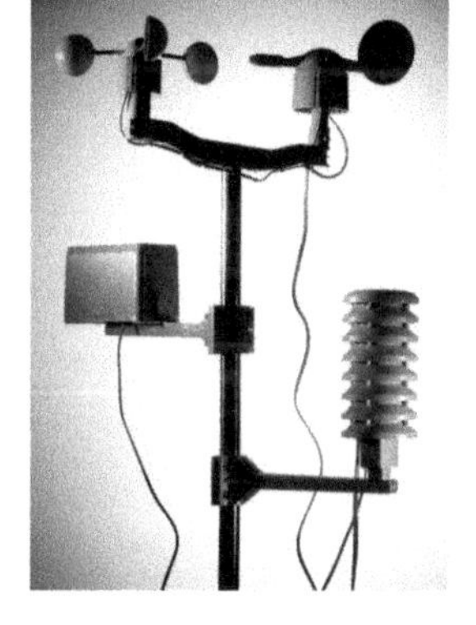

▲ **Figs. 16.2 and 16.3** *Digital instruments have the benefit of giving very accurate results. Can you identify the instruments shown?*

Advantages of digital recording instruments

- They measure to a far finer measure than is possible on a non-digital instrument.
- The reading is obtained quickly and easily, as the data is expressed clearly in figures.
- The reading is more likely to be accurate, as it cuts out the chance of misreading by parallax errors, reading the wrong end of the index and other errors.
- Totals or averages for the day can be quickly calculated.
- Portable versions can be used at any site.
- Readings download directly to a computer, saving time in accessing the data.
- Digital readings are also very frequent, so more analysis can be done, such as calculating hourly averages.

Key ideas

- **Clouds consist of tiny water droplets or ice particles which are too light to fall to Earth.**
- **Clouds with the greatest vertical extents form in the tropical zone where the tropopause is at its highest.**
- **Clouds are classified according to their height and shape.**
- **High clouds (above 6 km from the surface) have the prefix *cirro-* and medium level clouds (2–6 km above the surface) are prefixed *alto-*.**
- **Layer-shaped clouds are described as stratus or have the prefix *strato-*, whereas globular cloud is cumulus or prefixed *cumulo-*.**

Cloud types and extent

How clouds form

Clouds form when moist air rises and cools, until water vapour condenses around condensation nuclei into water droplets or, if it is sufficiently cold, to ice crystals. Air continues to rise while it is warmer and lighter than the air into which it is rising. It cannot rise above the tropopause.

Clouds only produce precipitation if they have a lot of water or ice particles, so that the particles can collide and join together to become heavy enough to fall through the rising air currents. The only clouds that produce heavy precipitation are **nimbostratus** and **cumulonimbus**. Only drizzle, fine rain, falls from stratus.

The main types of cloud

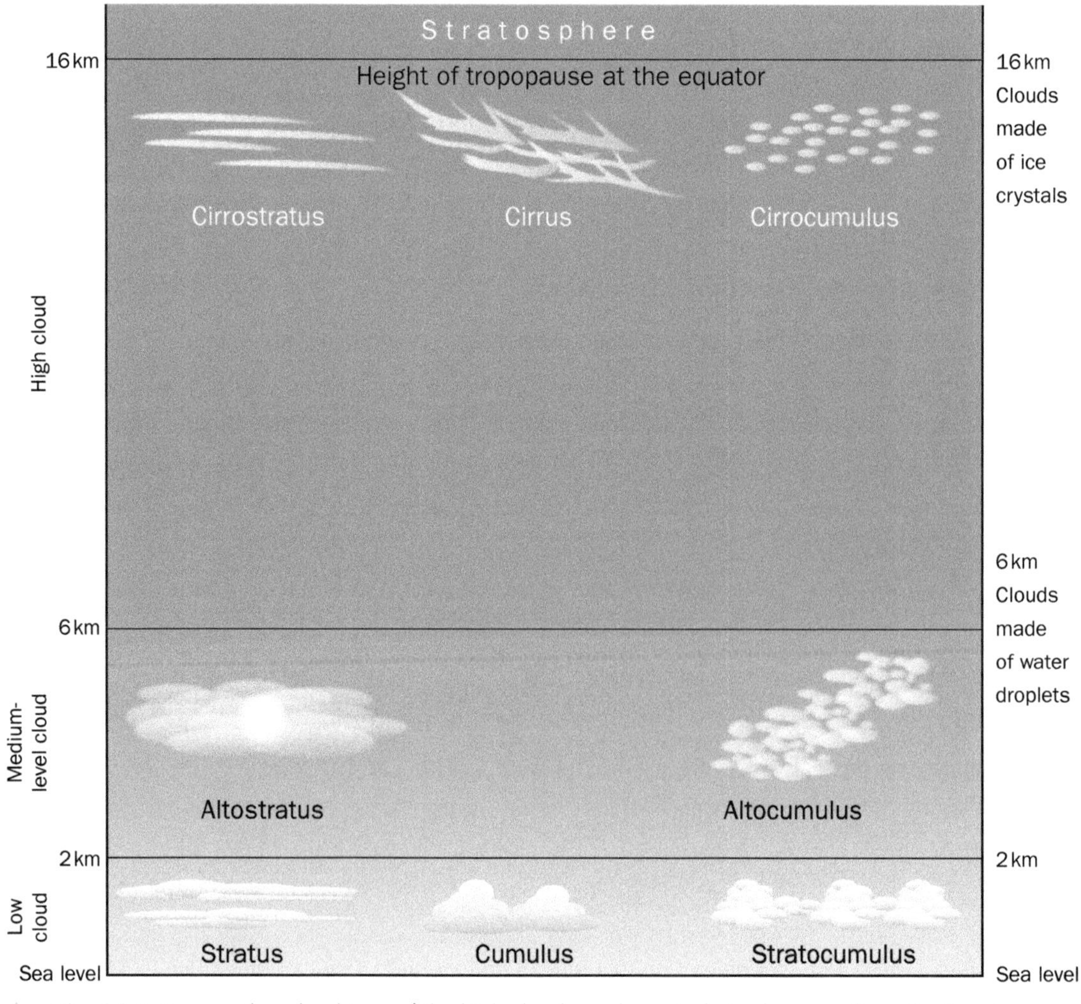

Fig. 16.4 *Fair weather cloud types. (The high clouds are lower in latitudes away from the Equator, where the tropopause is lower.)*

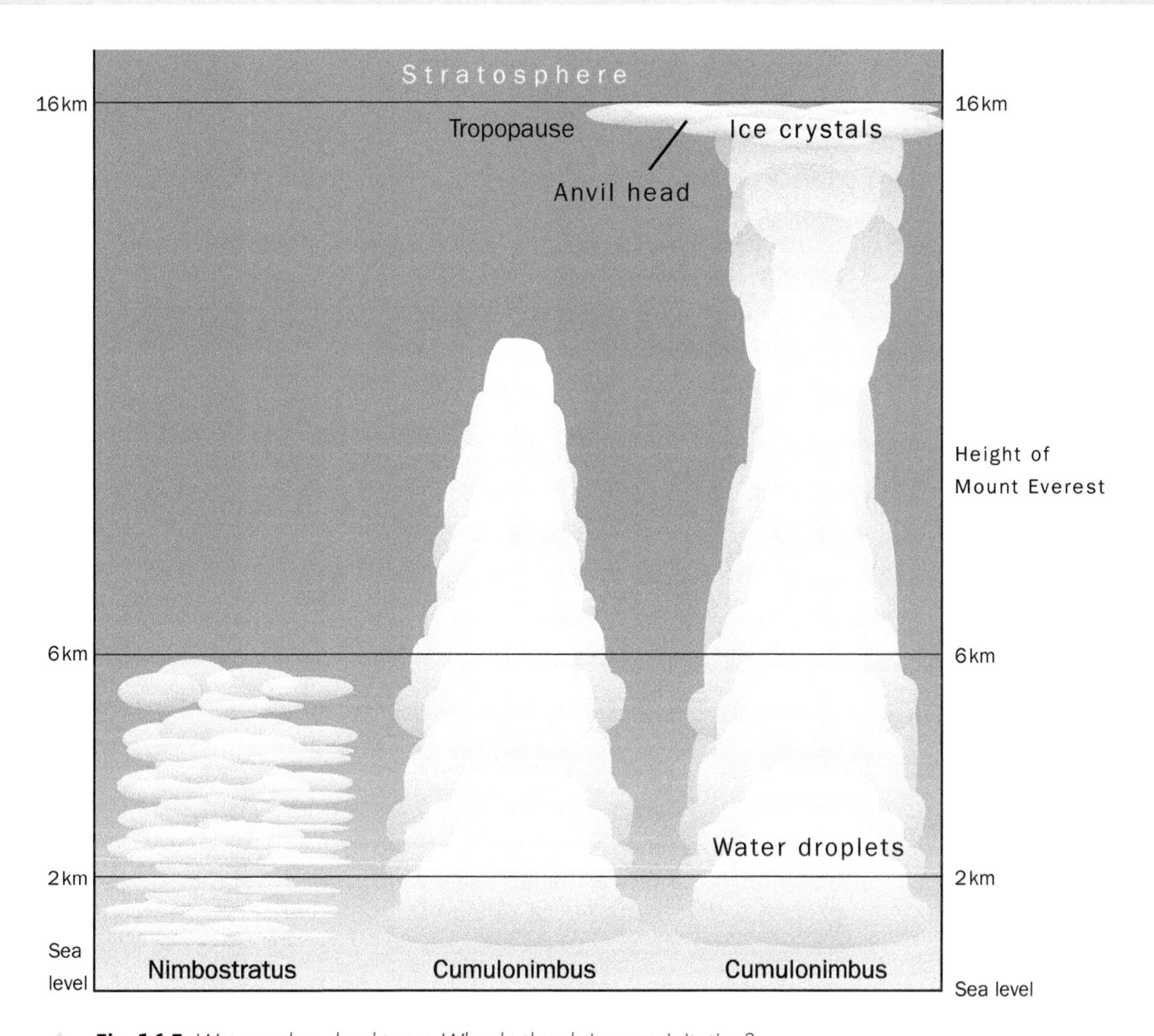

Fig. 16.5 *Wet weather cloud types. Why do they bring precipitation?*

How cloud extent is measured

The extent of cloud cover is estimated by eye, looking directly overhead, and expressed in the number of oktas (eighths) of the sky covered with cloud.

CLOUD

Symbol	Cloud amount (oktas)
○	0
⦶	1 or less
◔	2
◔	3
◑	4
◑	5
◕	6
◕	7
●	8

Fig. 16.6 *Symbols used to describe cloud cover in oktas*

Recap

Weather is recorded by instruments which can be either manually read or are digital with an automatic continuous download to a computer. Both types have certain site requirements for accurate results. Most have to be in the open away from shelter or shade, but thermometers are housed in a Stevenson screen as shade temperatures must be recorded. Readings from a manually read instrument is from a scale or dial.

Manually read instruments at a weather station should be read at the same time each day. The eye must be level with the meniscus when reading from a thermometer or measuring cylinder. If the thermometer has an index, the reading is taken from the part of the index nearest the mercury.

Cloud extent, given in oktas, is the only weather reading that is estimated by observation.

Apply

See www.oxfordsecondary.com/esg-for-caie-igcse for the answers to the 'Apply' task.

Describe how the following manually read instruments work and how you would obtain an accurate reading from them: maximum and minimum thermometer; hygrometer; wind vane; anemometer; rain gauge; sunshine recorder; barometer.

Review

Digital instruments have the following advantages:

- clear numerical displays that are quickly read and accurate
- portable so can be taken from site-to-site for fieldwork purposes
- readings frequently transmitted to a computer saves time and allows more analysis, such as the calculation of hourly averages. It is particularly useful for readings of wind speed, which are very variable.

Cloud types are identified according to their shape and height.

You should be able to label diagrams to show the features of the weather instruments. There are photographs and detailed descriptions of the manually read weather instruments and of each cloud type in Chapter 6 of *Complete Geography for Cambridge IGCSE and O Level.*

Sample question

1. You are setting up a weather station in the school grounds. Name the traditional instruments you would buy and explain how they should be sited in order to provide accurate records. **[7 marks]**

Analysis

- ✓ This question tests your knowledge of the special requirements necessary for the siting of weather instruments in order to give accurate results.
- ✓ To achieve a good mark, you need to think broadly about the variety of instruments required. It is best to try to give at least one site requirement for each instrument.
- ✓ To score the highest mark you need to note that the task states 'to provide accurate records' and that this is best done by explaining how the site requirements enable accuracy to be achieved.
- ✓ You should answer every part of the question for at least two of the instruments.

Student answer

The rain gauge should be away from trees to avoid drip adding extra to the rain and away from buildings to avoid shelter, which would give smaller readings than the actual rainfall. The wind vane and anemometer need to be placed at a height well above any obstructions that could shelter them. The hygrometer and Six's thermometers would be placed in a Stevenson screen to give shade readings at a standard height from the ground, away from radiation of its heat or cooling at night and away from buildings that radiate heat. The sunshine recorder should be in an open location, facing south in the Northern Hemisphere outside the tropics or north in the Southern Hemisphere. The barometer needs to be out of sunlight which would give a lower than correct air pressure because warmed air expands.

[Level 3: 7 out of 7]

See www.oxfordsecondary.com/esg-for-caie-igcse for the mark scheme for this question.

Examiner feedback

This answer scores seven marks, because it states at least one site requirement for each instrument and explains the reason for it in the cases of the rain gauge (to avoid extra rain or too little being recorded) and for the barometer (to avoid a lower than correct pressure reading).

Try to think what could have been stated for the other instruments to explain why each of them would give an accurate reading if the site requirements given were used.

Key ideas

- Places with hottest temperatures in the middle of the year are in the Northern Hemisphere.
- Places with hottest temperatures at the beginning and end of the year are in the Southern Hemisphere.

Common error

Be aware that ranges of rainfall are not normally calculated for climate description, but ranges of temperature are used.

Weather data calculations and descriptions

Rainfall

Know how to calculate:

- monthly and annual rainfall totals
- mean (average) monthly and annual rainfall, over a minimum of thirty years.

Annual rainfall amounts can be described as in the table below.

Annual rainfall (mm)	Description of the amount
0–249	Very low
250–499	Low
500–999	Moderate
1000–1999	High
Over 2000	Very high

Temperature

Know how to calculate:

- daily (diurnal) range of temperature
- mean daily temperature
- mean monthly temperature
- annual range
- mean annual temperature.

Temperature and temperature ranges can be described as in the tables opposite.

Temperature (°C)	Description
30 and above	Very hot
20–29	Hot
10–19	Warm
0–9	Cool
-10 to -1	Cold
Below -10	Very cold

Temperature range (°C)	Description
0–3	Very small
4–8	Small
9–19	Moderate
20 and above	Large

Pressure

A higher pressure area surrounded by lower pressure is a high pressure system. Low pressure areas are surrounded by higher pressures.

Wind

Wind speed can be described as in the following table.

Wind speed (km/hour)	Description of the wind speed
Below 20	Calm or light
20–50	Moderate to strong
50–100	Gale
101–118	Storm
119 and above	Hurricane, typhoon or cyclone

Graphs and diagrams drawn from weather recordings

Graph	Description
Climate graph	These show mean monthly temperatures and mean monthly precipitation calculated over at least 30 years.
Dispersion diagrams (see also page 171)	These are useful for showing distributions. The range, median (middle number) and modal (most frequently occurring) values can be quickly read and compared on them. The position of the calculated mean or median value, whichever is more appropriate, can be shown on the diagram.
Wind roses (see also page 170)	These have bars radiating from a central area for each compass direction. The length of each bar shows the number of days with the wind from that direction over a period of time (usually a month). The number of calm days is written in the centre.
Choropleth maps (see Fig. 5.3, page 31)	Conventional shading should be used: heaviest for the largest value and progressively lighter with decreasing values.
Weather maps or synoptic charts (see Question 1, Chapter 17)	Readings taken at meteorological stations at a certain time are plotted on synoptic charts, using sets of symbols to show the weather at weather stations across the area.

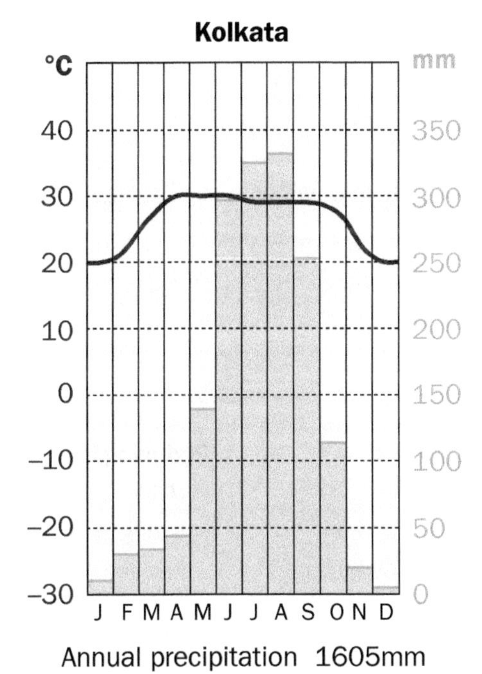

Fig. 17.1 *Climate graph for Kolkata*

Common error

Remember that summer is the hottest season (high sun period) and winter the coldest (low sun period). Be aware that the months differ in each hemisphere.

Common error

Remember to put the units (mm, °C, mb, %, etc.) when reading off information from a weather or climate diagram.

If plotting a temperature, be careful to position it accurately. Use the scale to see if it should be on a line or in a space. Remember to join the temperature plots with a line to the line already plotted.

Keep the following points in mind when describing climate graphs.

- Equatorial climates have no seasons, but refer to summer and winter when describing climate graphs of other locations.
- Do not give a month-by-month account of the changes in temperature, but quote figures and describe in words the highest and lowest temperatures, the mean and the range of temperatures.
- Remember to describe both the total and seasonal distribution of the rainfall.

Key ideas

- Tropical storms form over warm tropical oceans between May and November in the Northern Hemisphere and between November and May in the Southern Hemisphere.
- They move pole-wards and westwards to strike east coasts of continents in the tropics and sub-tropics, but do not form near the Equator. (Mexico has the only west coast that is affected.)
- They have winds of at least 101 km/h.
- Droughts can occur almost everywhere but some areas have particularly severe and frequent droughts, such as in the Sahel of Africa.
- The effects of droughts are worse in LEDCs than in MEDCs.
- The problems caused by droughts are worse when the rains fail in successive years.
- Droughts cause natural vegetation and crops to die, giving problems for farmers and people.

Weather hazards

You will not be required to explain the causes of these hazards, but you will need to know the impacts of tropical storms and desertification.

Tropical storms

The strongest tropical storms with winds of at least 119 km/h are known as:

- cyclones in the Indian Ocean, Bay of Bengal and northern Australia
- typhoons in the South China Sea and west Pacific Ocean
- hurricanes in the Caribbean Sea, Gulf of Mexico and west coast of Mexico.

Effects of tropical storms	Reducing the risk from tropical storms
• Loss of life. • Ferocious winds destroy buildings, power lines and communications. • Heavy rainfall causes flooding, landslides and mudflows. • Storm surges greatly damage coasts. • Trees are uprooted and crops destroyed. • Ships are wrecked and sunk. • Effects are greatest over heavily populated areas. • Costs are highest in MEDCs where property is of higher quality • Destruction is most in LEDCs which are less well prepared with flimsier structures.	• Evacuate people from areas in danger. • Satellite images track storms to provide warning, although predictions of where the storm will hit are not always successful because they can suddenly change course. • Construct seawalls and artificial levees to try to prevent flooding. • Construct buildings to withstand strong winds. • Store emergency supplies of food and water in emergency shelters with toilets. • Board windows up. • Insure properties.

Drought and desertification

The term *desertification* used to be used only for land degradation caused by the adverse impact of human activities (Chapter 21). However, it now includes natural changes in soil and vegetation caused by **drought** and climate change.

Common error

Remember that people who live in climates with a dry season or in a desert have adapted to the lack of rain. This is different from drought.

Consequences of drought

- The death of vegetation leaves the soil bare and soil erosion occurs, eventually causing desertification
- Death of pastures, leading to the death of livestock
- Crop failure and reduced crop yields, leading to rising world food prices
- Insufficient food and water for people and their animals
- Water supplies dry up, leading to conflicts between agricultural and domestic use
- Dust storms
- Malnutrition and sickness, especially in young children as food supplies reduce
- Death of wild animals and birds, as their habitats and food supplies reduce
- Bush fires, which destroy crops, vegetation, habitats and homes
- Population migrate to seek emergency aid, or permanently migrate to towns and cities.

Aid agencies

Their work

- Providing emergency medical aid, tents, food and water
- Drilling boreholes and wells, building dams and laying water pipelines
- Providing emergency farming kits, such as drip irrigation systems and water tanks.

Recap

Climate is the average of the weather over at least 30 years. Climates are distinguished by differences in temperatures and rainfall. Mean daily, monthly and annual temperatures can be calculated, as well as daily and annual temperature ranges. Monthly and annual rainfall totals and averages are also used to describe climates. The prevailing wind direction influences both temperature and rainfall.

Climate graphs show monthly averages of temperature and rainfall. Synoptic charts, isoline and choropleth maps illustrate variations over space. Wind roses and dispersion diagrams are also useful climatic diagrams.

If a climate has seasonal differences, the hottest time of the year (summer) will be in the middle of the year in the Northern Hemisphere, and at the beginning and end of the year in the Southern Hemisphere.

Tropical storms and droughts are particularly hazardous for people.

Apply

Name a destructive tropical storm that caused damage to a named coastal area. Explain why the tropical storm was so destructive and describe the nature of the damage it caused.

See www.oxfordsecondary.com/esg-for-caie-igcse for the answer to the 'Apply' task.

Review

Drought, common in tropical and temperate semi-arid climates, is a period of unusually dry weather that makes desertification and soil erosion worse. Desertification occurs when people misuse the land by using poor farming techniques, particularly by over-grazing and over-cultivating. Soil erosion then makes formerly productive areas no longer productive. When drought occurs in such areas, soil erosion and desertification occur even faster.

- Drought causes the death of any remaining vegetation, pasture and crops leaving the soil exposed to erosional processes.
- When vegetation no longer rots into the soil, there is nothing to stick the soil particles together and they become loose.
- When wind blows over bare, loose soil it is easily blown away in a dust storm.
- When heavy rain falls on bare soil there are no roots to anchor the soil and it is washed away. On slopes this cuts deep gullies into the landscape. On gentle slopes a sheet of water removes the surface layer of the soil.

Sample question

1. Study the synoptic chart for part of Africa at 2 p.m. on a day in March.

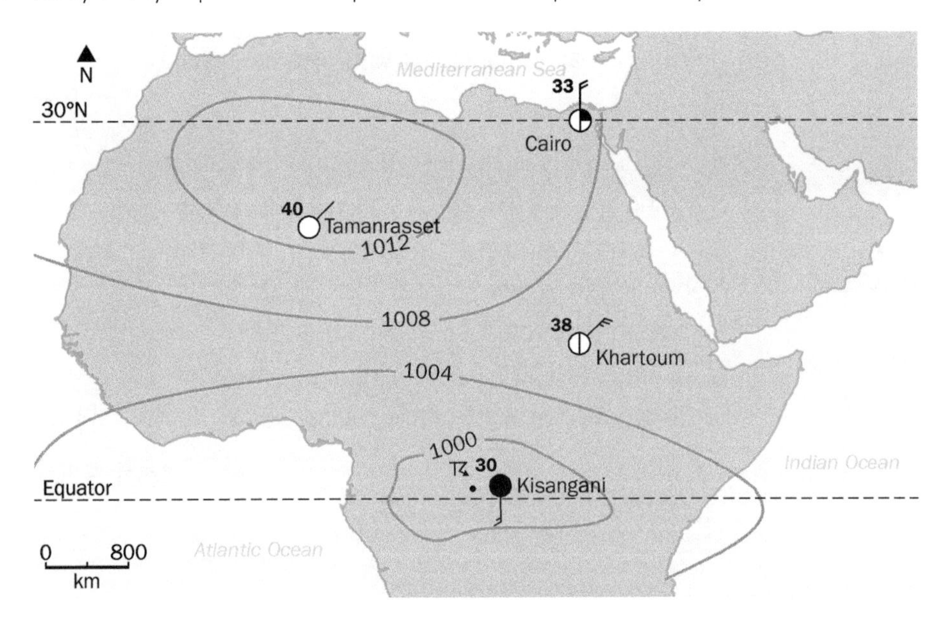

CLOUD

Symbol	Cloud amount (oktas)
○	0
⏀	1 or less
◔	2
	3
◑	4
	5
◕	6
	7
●	8

Wind speed (km/h)	Weather
Calm	Drizzle
1–5	Rain
6–15	Thunderstorm
16–25	
26–34	

For each additional half-feather add 10 km/h

90–98

(a) Identify the weather station with 1 okta or less of cloud cover. **[1 mark]**

(b) Compare the air pressure at Tamanrasset with that at Kisangani. **[2 marks]**

(c) Describe and explain the atmospheric temperature at Tamanrasset at 2 p.m. on the day shown in the synoptic chart. **[4 marks]**

Analysis

✓ This resource illustrates typical hot desert weather and how it differs from equatorial weather at Kisangani.

✓ The question requires knowledge of influences on insolation and of the transfer of heat from the ground to the air by conduction.

✓ It allows more practice of writing a comparison.

See www.oxfordsecondary.com/esg-for-caie-igcse for the mark scheme for this question.

Student answer

(a) Khartoum [1 mark]

(b) It is higher at Tamanrasset with 1012 compared with 1000. [1 mark]

(c) It is 40 °C. This is because the sun's insolation reaching the Earth's surface is very strong as there is no cloud cover to absorb or reflect it. Also, the sun is overhead at the Equator in March and so would be at a high angle at Tamanrasset at 2 p.m., giving concentrated heating power at the Earth's surface. [3 marks]

Total: 5 out of 7

Examiner feedback

Part (b) only scores one mark because the units of pressure have not been stated and the pressure would be more accurately described as *more than* 1012 mb and *less than* 1016 mb at Tamanrasset, compared with less than 1000 mb but more than 996 mb at Kisangani. (There are no 1016 mb and 996 mb isobars on the map.)

Part (c) is a good explanation, but the mark for the description would not be gained, because the figure is stated but not described.

Sample question

2. Study the temperature and rainfall statistics for a weather station.

Month	J	F	M	A	M	J	J	A	S	O	N	D
Average temp (°C)	19.0	18.5	18.0	17.0	16.0	15.5	14.5	14.0	15.0	16.0	17.0	18.0
Average rainfall (mm)	29	31	30	20	9	2	2	8	4	5	14	11

(a) In which hemisphere is this weather station and how does the graph show this? **[1 mark]**

(b) Describe the annual total rainfall and its seasonal distribution. **[2 marks]**

(c) Look at the map in Question 1. Identify ways in which the weather at Kisangani at 2 p.m. on that day in March was typical of daily weather in the equatorial climate. **[3 marks]**

Analysis

✓ The first question tests your knowledge that the sun is overhead in the Southern Hemisphere in December and so it is hottest (summer) at the beginning and end of the year.

✓ The second question is a reminder to read the question and do what it asks, that is state the season, not the months.

✓ The third question requires you to recognize that the daily weather shown on the map for a day in March is typical of every day in the equatorial climate, which is the same all through the year.

Student answer

(a) Southern, because it is hot in summer. [0 marks]

(b) The weather station has a very low annual total rainfall, with hardly any rainfall from May to October – so little it can be described as dry and it has more in January to March. [1 mark]

(c) It has a temperature of 30 °C, which is hot and it is raining. That is typical afternoon weather in the equatorial climate which is a hot, wet climate. [2 marks]

Total: 3 out of 5

Examiner feedback

In part (a), the candidate has named the correct hemisphere, but the reasoning is not acceptable. The answer must state that the *hottest* time of year, summer, is the beginning and end of the year.

In part (b), mention of months is irrelevant when the question asks for a *seasonal* description, so the season mark would not be gained, but the description of the total annual rainfall is accurate.

Two of the required three points have been given in part (c), together with an explanation of why they are typical of the climate. Two more points were possible: the sky is covered with cloud with 8 oktas and this is a cloudy climate; also in this climate the cloud builds up in the afternoon, often giving a storm, as shown on the map by the thunderstorm symbol.

See **www.oxfordsecondary.com/esg-for-caie-igcse** for the mark scheme for this question.

18 Climate and natural vegetation: Equatorial climate and tropical rainforest

Key ideas

- Equatorial climate and vegetation is found within 10° of the Equator.
- The climate results from low pressure in these latitudes.
- The climate is uniform with no seasons. It is hot and wet all year.
- Tropical rainforest is adapted to grow in the equatorial climate.

Characteristics of the equatorial climate

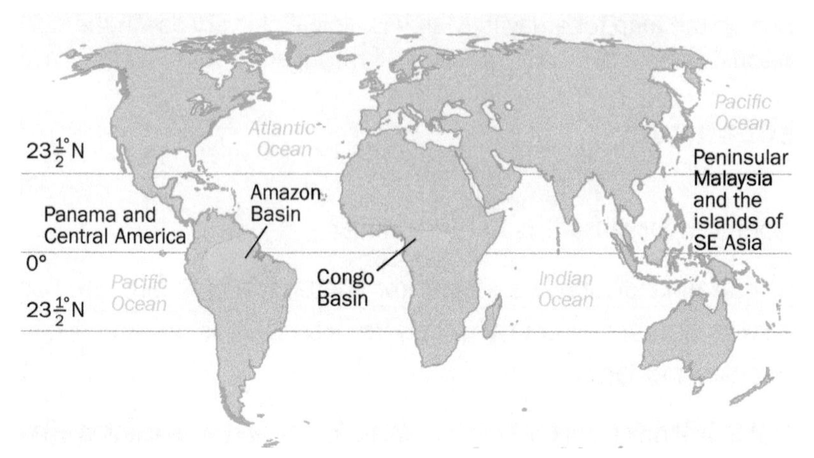

Fig. 18.1 *Areas with an equatorial climate and tropical rainforest*

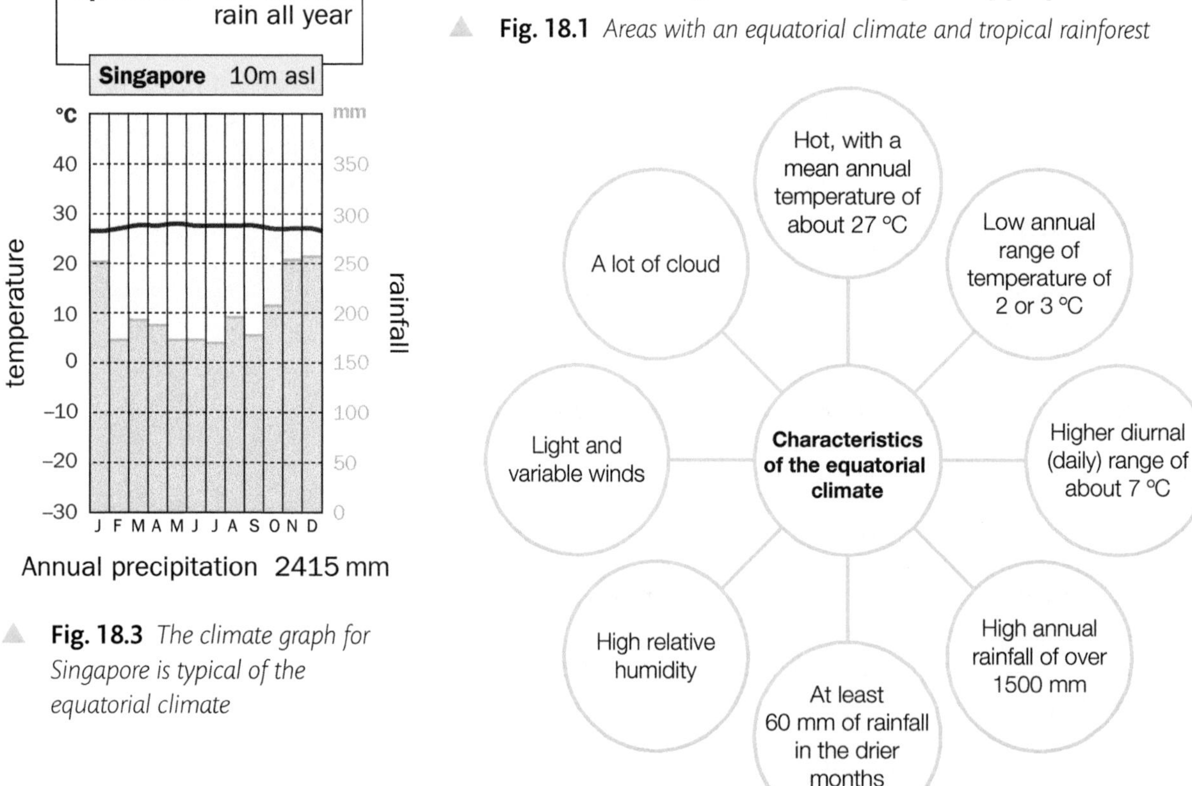

Fig. 18.3 *The climate graph for Singapore is typical of the equatorial climate*

Fig. 18.2 *You will need to know these characteristics of the equatorial climate*

Influences on the temperatures

- The sun has high heating power because it is overhead at the Equator on 21 March and 23 September, and high in the sky at midday all year in low latitudes.
- Cloud reduces the daytime temperature and keeps it warm at night by absorbing radiation from the Earth.

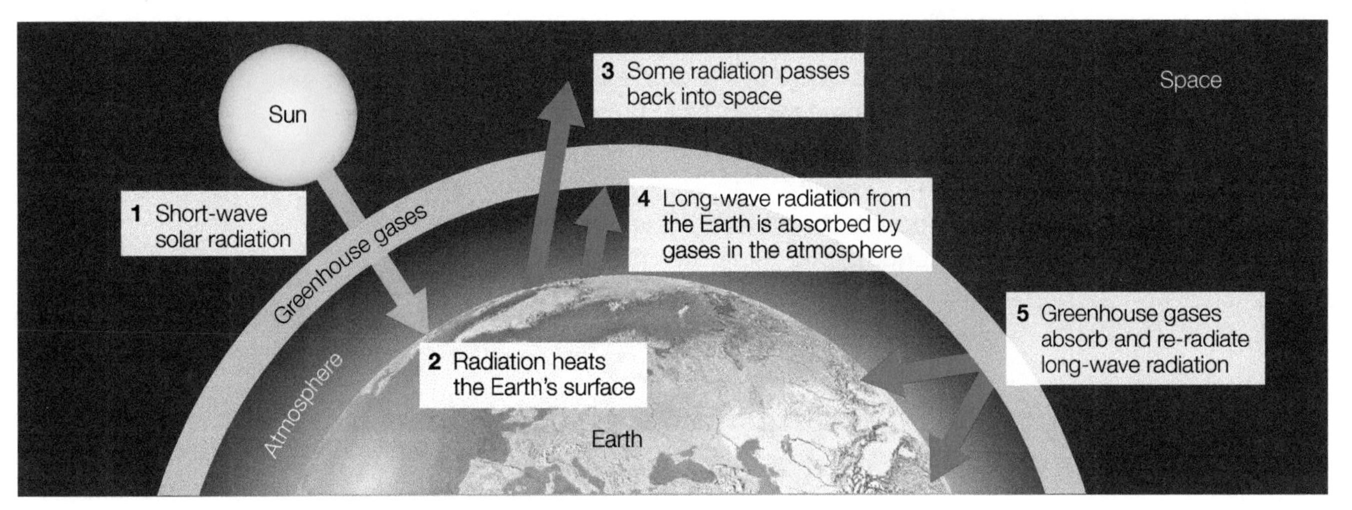

Fig. 18.4 *The greenhouse effect. Why is the name appropriate?*

Reasons for the high rainfall all year

Convection

The combination of high temperatures and of air with a high moisture content results in **convectional rainfall**.

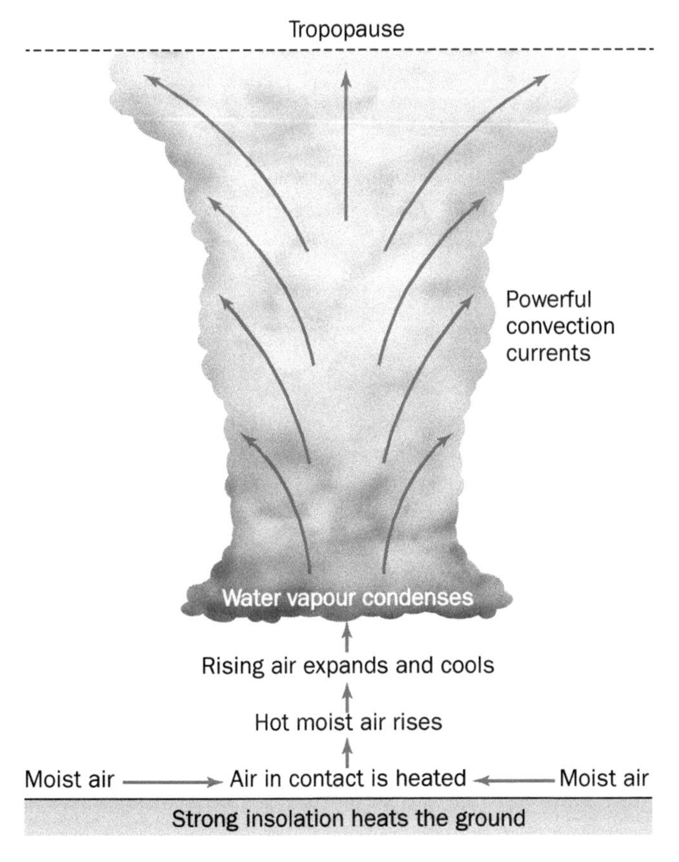

Fig. 18.5 *These towering cumulonimbus clouds produce very heavy rain, often with thunder and lightning*

As air is rising, surface winds are light and low pressure results. Also, warm air expands, giving less weight in a column of warm air than in colder air.

The equatorial low pressure system

The equatorial zone has permanent low atmospheric pressure. NE and SE trade winds blow into the low pressure belt to replace the rising air. They meet in the **Inter-Tropical Convergence Zone (ITCZ)** and rise, so cloud forms and heavy rain falls.

Common errors

When describing the greenhouse effect remember that it is *radiation* from the Earth not *reflected* rays that are absorbed by the atmosphere.

Never state that air condenses– it is water vapour in the air that condenses.

Key ideas

- **Tropical rainforests need a mean annual temperature of 24 °C and a minimum annual rainfall of 1500 mm.**
- **They are well adapted to the equatorial climate.**
- **Forests maintain soil fertility, water quality and help to keep the climate stable.**
- **The tropical rainforest has a very rich and diverse animal life.**
- **The effects of deforestation on the natural environment are both local and global.**

Tropical rainforest ecosystems

Tropical rainforest vegetation

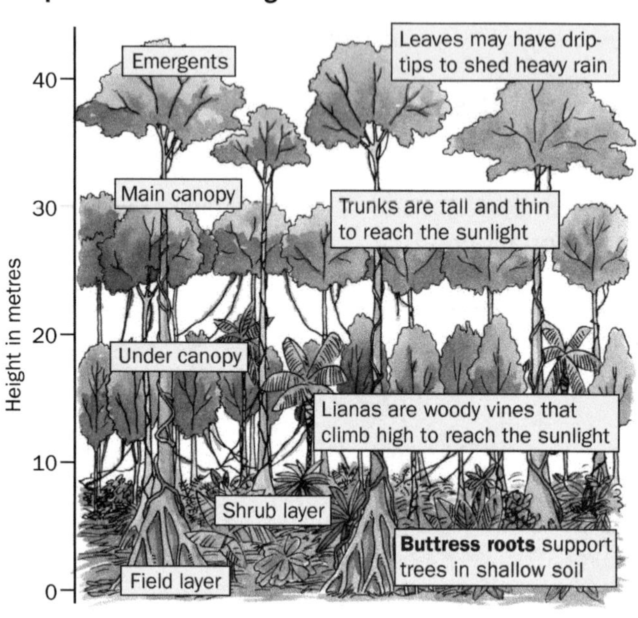

Fig. 18.6 *The forest has a five-tier structure*

Other characteristics:

- There are a great number of tree species in an area but they all look alike.
- Each species is widely spread apart.
- They are mainly hardwoods, such as ironwood and mahogany.
- Rainforests are so dense and continuous that light does not penetrate far.

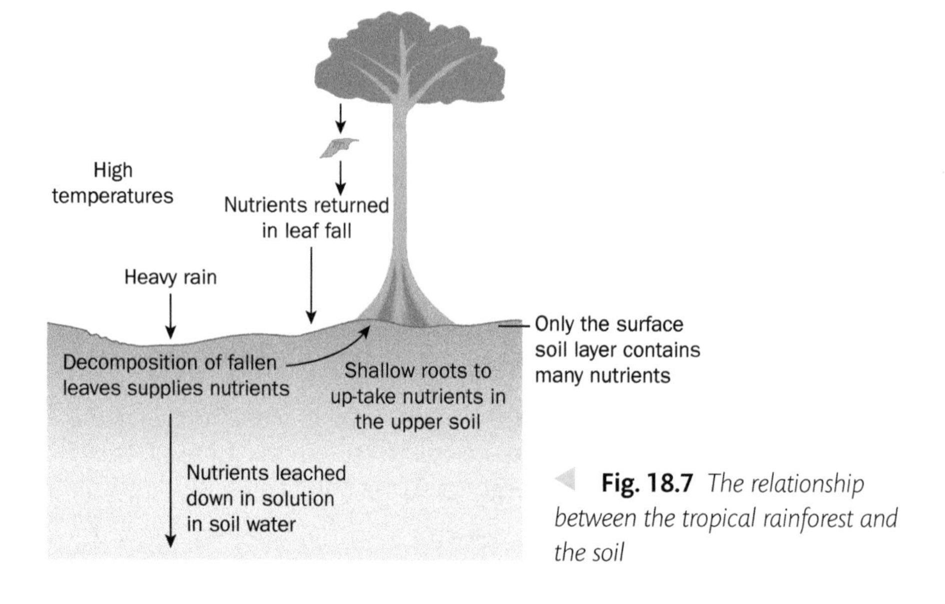

Fig. 18.7 *The relationship between the tropical rainforest and the soil*

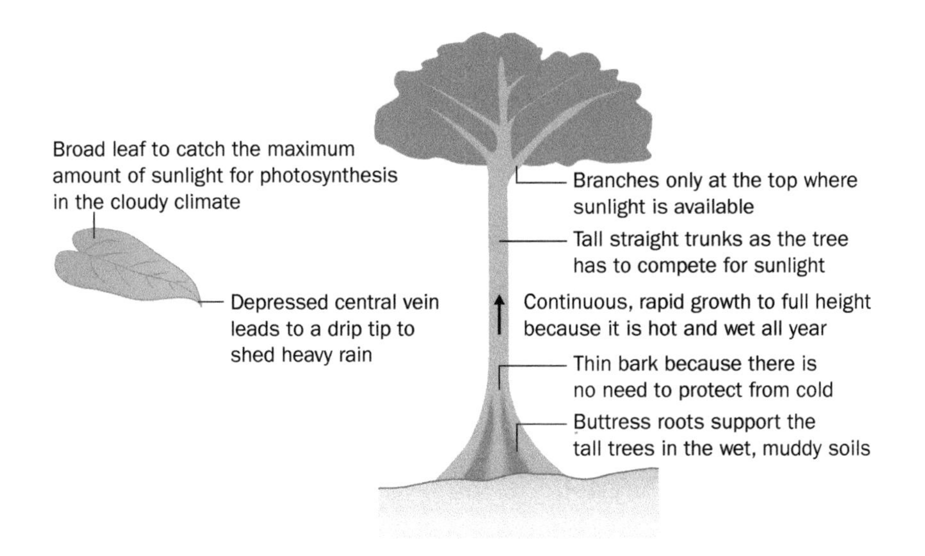

Fig. 18.8 *The relationship between the natural vegetation and the climate*

The forest is not seasonal because the climate has no seasons. A tree may have branches with no leaves while others have full foliage.

Common error

Remember that soil fertility has nothing to do with how much rainfall is available for plant growth. It is simply about its nutrient status.

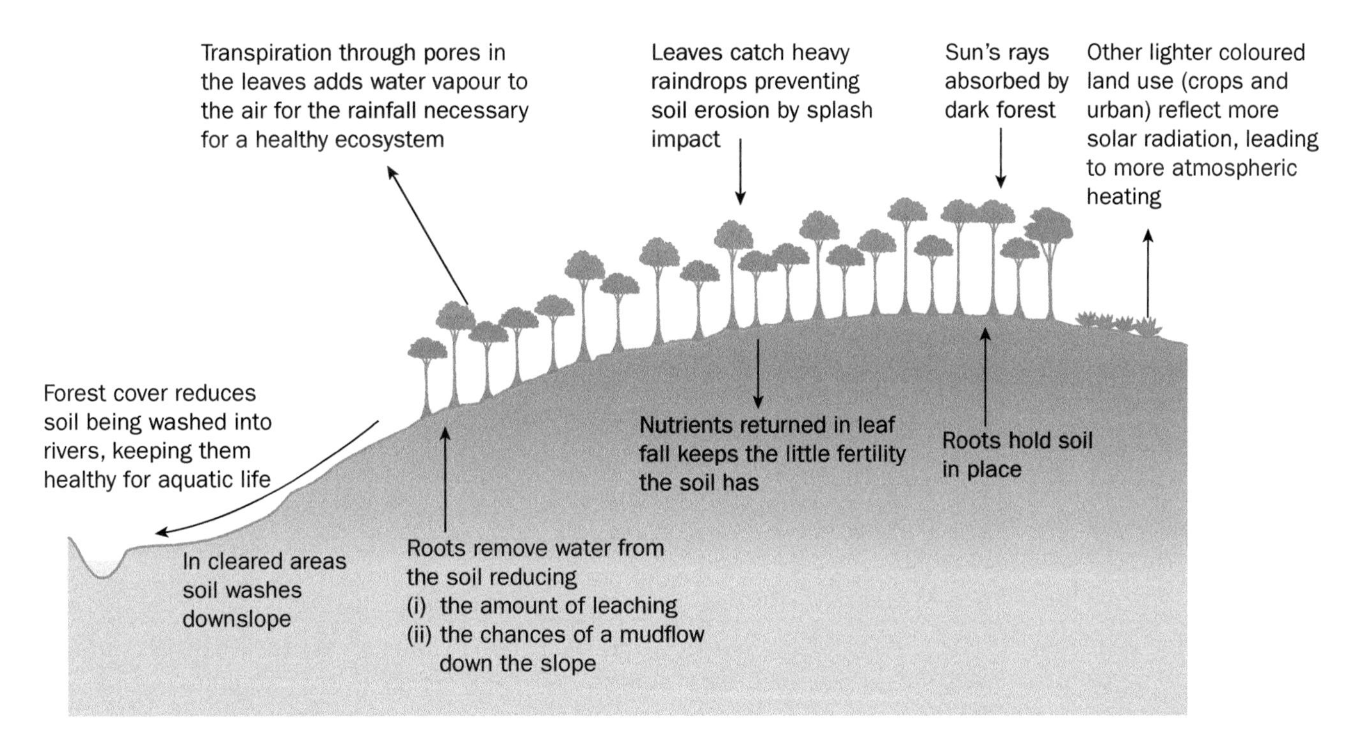

Fig. 18.9 *Tropical rainforests maintain a healthy ecosystem*

Exam tip

Be able to describe and explain a case study of an area of tropical rainforest. Know about its location, climate, vegetation and ecosystem, as well as the causes and impacts of deforestation.

Causes of deforestation

- Multinational companies mine raw materials.
- Some establish commercial pastoral farms (cattle ranches) and plantations (large commercial crop farms), such as oil palm.
- Hydroelectric power stations are developed.
- Logging for hardwoods, sometimes without re-planting.
- Roads built to access these developments open up areas for further deforestation.
- Population pressure encourages felling for subsistence farming and fuel.

Consequences of deforestation

- Clearance by burning has covered far-away countries in smoke and ash.
- Burning forests emits a lot of carbon dioxide, contributing to enhanced global warming.
- Local reduction in rainfall because of reduced transpiration.
- Loss of plant species which may have unknown uses
- Many forest animals are now endangered species.
- Loss of habitats and animals, such as the orangutan in Borneo.
- Soil erosion and loss of soil fertility.

The equatorial climate is hot and wet all year, so its natural vegetation grows rapidly and continuously all year.

The tropical rainforest is the richest ecosystem on the Earth with the richest, densest, tallest vegetation. It has the greatest diversity of plants and animals because it provides a variety of habitats and an abundance of food for animals.

Each layer of the forest has different conditions of sunlight, temperature and humidity to suit different plant and animal life.

Explain why tropical rainforest is beneficial both for the atmosphere and the wildlife that inhabit it.

See **www.oxfordsecondary.com/esg-for-caie-igcse** for the answer to the 'Apply' task.

Review

Plants take in carbon dioxide, a greenhouse gas, so forests are 'carbon sinks', reducing the potential for global warming. They also respire oxygen, vital for life.

Trees recycle water as their roots take it in and their leaves release water vapour to the air.

Deforestation leads to a much poorer ecosystem, as nutrients are leached out of the reach of plant roots. Rapid soil erosion and soil degradation occurs. Plant regrowth is less rich than before and the climate becomes drier. Agricultural yields reduce as soil nutrients become exhausted. Fish die as rivers become choked with soil. Animals lose their food and habitat, so many are near extinction. Valuable sources of medicine and traditional ways of life of hunter-gatherers are disappearing.

The increase in greenhouse gases leads to more global warming. Careful management and conservation measures are necessary to preserve the forest for the sake of the planet, but it is too large to patrol effectively.

Detailed descriptions of animal life in the tropical rainforest are included in Chapter 6 of *Complete Geography for Cambridge IGCSE and O Level.*

Sample question

1. Statistics of mean monthly rainfall for a weather station near the Equator in Africa are shown below.

Jan	Feb	Mar	Apr	May	Jun	Jul	Aug	Sep	Oct	Nov	Dec
60	80	175	159	132	104	136	169	186	121	195	83

(a) Explain why the air in equatorial climates is very moist. **[1 mark]**

(b) Describe how the rainfall at this place is typical of an equatorial climate. **[2 marks]**

(c) Explain why convectional rain usually falls in the afternoon in areas with an equatorial climate. **[4 marks]**

Analysis

✓ The equatorial climate is one of the two climates that you need to learn.

✓ You need to be able to describe figures given either in a data table, as here, or shown on a graph.

✓ You should be able to explain characteristics of the climate.

✓ The command term 'explain' means you must give reasons for.

✓ The command term 'describe how' means you should use the data given to state why that indicates an equatorial climate.

Student answer

(a) The air is moist because it has a high temperature. [0 marks]

(b) There is a high rainfall total, with at least 50 mm in each of the three drier months. [2 marks]

(c) By the afternoon, the sun has been heating the land surface for hours, as there is little cloud in the mornings and also putting lots of moisture into the air by evaporation from rivers and lakes. The air next to the Earth has been heated, which causes it to expand and rise. As it rises, it cools and condenses, giving torrential rainfall. [4 marks]

Total: 6 out of 7

Examiner feedback

The answer to part (a) has a correct idea, but it has not been used to *'explain'*, the command term for this question. The answer needs to include a reason why hot, equatorial air becomes moist. Two correct observations have been made in the answer to part (b). It might also be noticed that there are two periods of peak rainfall – in March and September, when there will be overhead sun at the equinoxes, causing an increase in the convectional rainfall. There is an error in part (c), which is otherwise a good account: *air* does not condense. It is the *water vapour* in the air that condenses.

See www.oxfordsecondary.com/esg-for-caie-igcse for the mark scheme for this question.

Key ideas

- Most tropical deserts are located astride the tropics, between latitudes 15° and 30° on the western sides of continents.
- The Mojave and Thar Deserts lie between the Tropic of Cancer and 40 °N.
- The climate results from high pressure in these latitudes.
- Tropical desert climates are hot and dry all year.

Characteristics of the hot desert climate

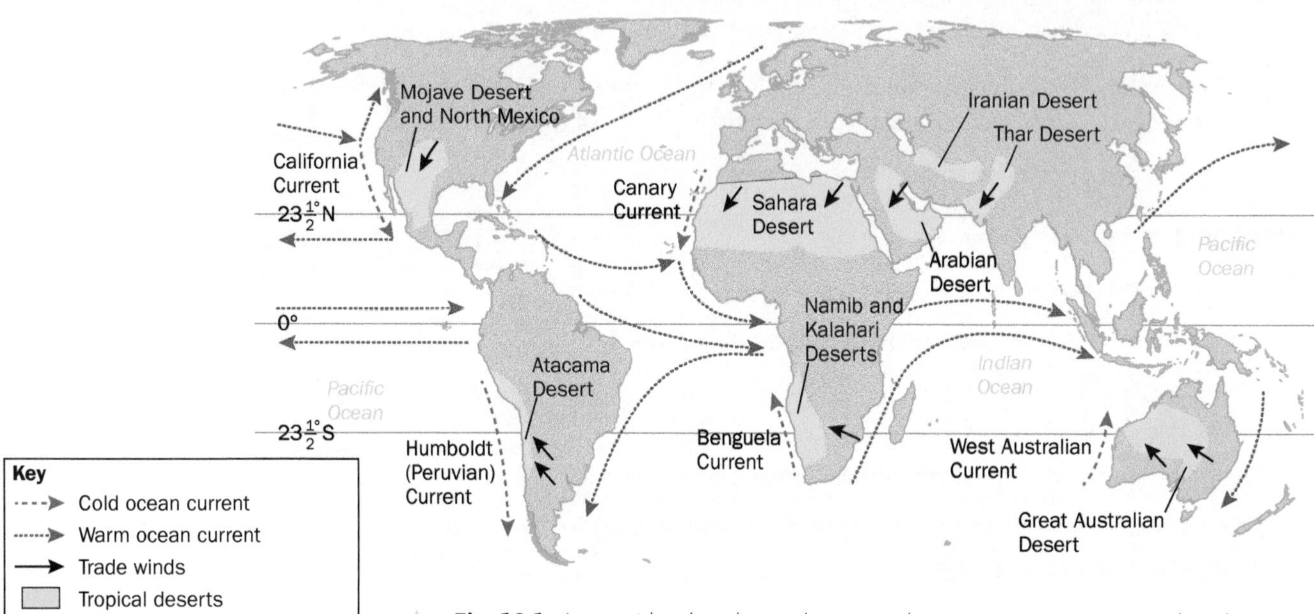

Fig. 19.1 *Areas with a hot desert climate and vegetation. Can you see what their locations have in common?*

- Very hot summers and hot or warm winters
- Moderate annual temperature ranges and very high daily temperature ranges
- Very low rainfall – below 250 mm with occasional torrential rain
- Dry NE trade winds in the Northern Hemisphere and SE trades in the Southern Hemisphere
- High sunshine totals, low relative humidity and high pressure.

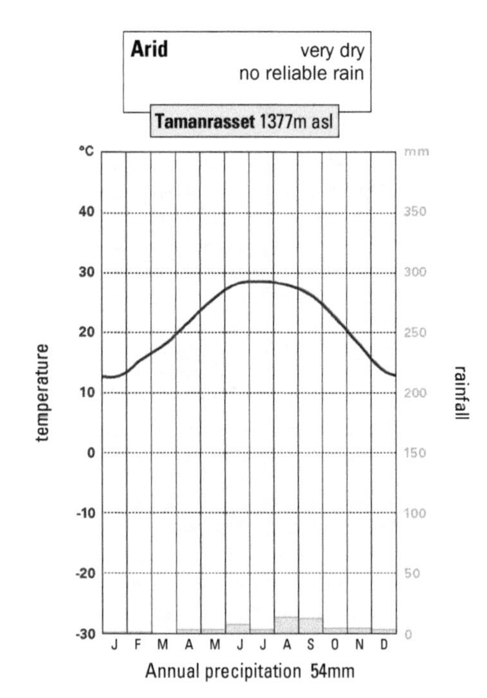

Fig. 19.2 *The climate graph for Tamanrasset in Algeria is typical of the hot desert climate*

Influences on the temperatures

- Latitude: summers are very hot because the sun is at a very high angle and winters are hot because the noonday sun is never very low at these latitudes.

Date	Location of overhead sun at noon	Season in Northern Hemisphere desert	Season in Southern Hemisphere desert
21 June	23½°N	Summer	Winter
22 December	23½°S	Winter	Summer

- Air temperature decreases as **altitude** increases.
- Coastal areas have warmer winters and cooler summers than inland places, so annual temperature ranges are smaller than inland.
- Cold offshore ocean currents lower the temperatures because onshore winds are chilled by contact with them.
- Lack of cloud causes extreme temperatures and very large daily temperature ranges.

Reasons for the low rainfall all year

Common error

It is not an acceptable explanation of the low rainfall in a hot desert to write that it is because the air is 'dry'. The reasons for this must be stated.

1. High pressure

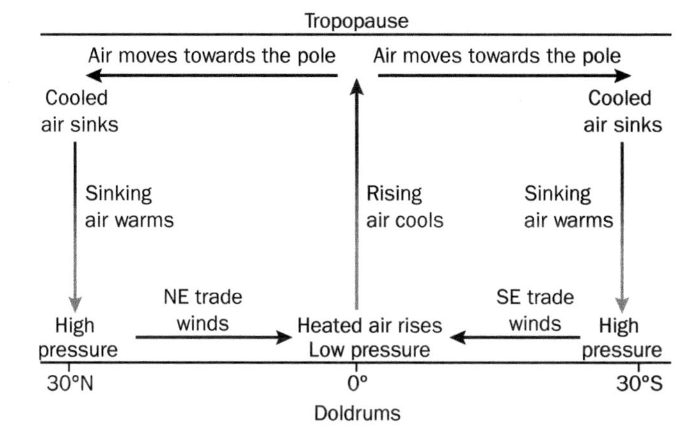

Fig. 19.3 *The circulation of air causing high rainfall in equatorial climates and low rainfall in hot deserts*

The trade winds are strong, constant and dry. Sinking air and dry offshore trade winds cause the very low precipitation of the deserts. The directions of the trade winds result from:

- winds blowing out of the high pressure areas into the equatorial low pressure zone
- winds being deflected by the Earth's rotation to the right in the Northern Hemisphere and to the left in the Southern Hemisphere.

2. Cold ocean currents offshore

Coastal deserts have condensation over the cold current, forming fog. This removes moisture from air moving inland.

3. Relief

The **rain shadow** side of a mountain is dry.

4. Temperature

Occasional convectional storms occur in the summer heat.

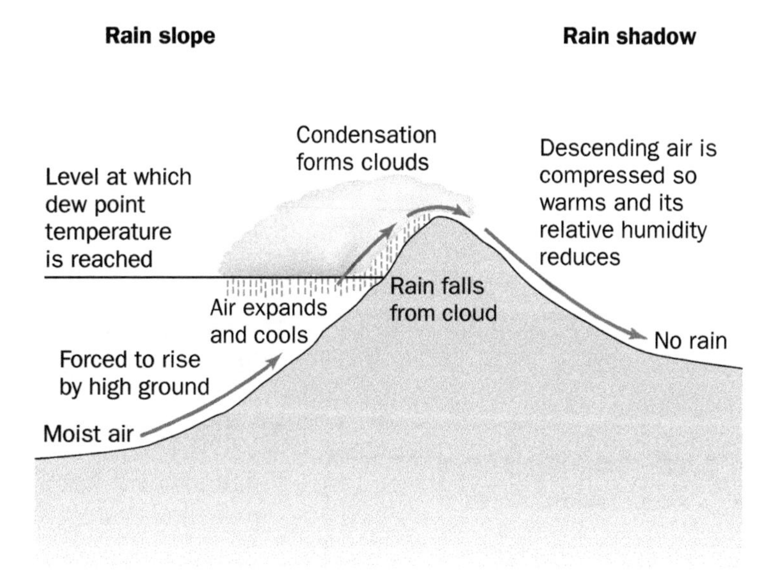

Fig. 19.4 *The influence of relief on rainfall*

Key ideas

- **Vegetation survives by being well-adapted to the lack of rainfall.**
- **The soils make it difficult for plant life to thrive.**
- **Desert animals have adaptations to the hot, dry conditions to enable them to survive.**

The tropical desert ecosystem

The vegetation in a tropical desert is adapted to its climate and soils. Leaves reduce water loss by having small surface areas with few **stomata** through which transpiration can occur. After rain, plants complete their lifecycles very quickly. Some have shallow, wide-spreading roots to catch rain before it evaporates.

Fig. 19.5 *Some plant adaptations to the dry desert climate*

Desert soils cause difficulties for plants

- Being rocky, stony or sandy, soils are very **porous** so water passes rapidly into them after rain.
- Sandy soils are mobile, so plants can easily be covered, and are also loose, so plants can be uprooted.
- Desert soils are thin and contain very few plant nutrients, as very little **organic matter** is available to decompose into them.
- Many desert soils are saline. Only salt-tolerant plants, such as saltbush, can grow in them.

The relationship in a hot desert ecosystem between wildlife and the soils, vegetation and climate

Adaptations to very dry conditions

- Zebras of the Namib Desert can detect pools of water below the surface with their nostrils.
- Fennec foxes in the Sahara and ostriches in the Namib and Kalahari Deserts can go for long periods without water. The foxes also reduce water loss by being nocturnal.
- Desert tortoises in the Mojave Desert can store moisture to last them a year.

Common error

It is rarely correct to describe vegetation as 'spread out' when what is meant is that it is sparse and scattered.

Adaptations to the wide range of temperatures

- Fennec foxes have thick fur which keeps their bodies cool in the hot day and warm in the cold night.
- Many desert animals have light-coloured fur or feathers to reflect the sun.
- Many desert animals are small and shelter in the daytime by staying in cool burrows. Some are nocturnal and hunt in the cool of the night.
- Horned vipers bury themselves in sand, partly to protect themselves from the heat, and partly for camouflage.
- Desert antelopes adapt to the seasonal temperature difference by having white summer coats and grey winter coats to stabilize their body temperatures.
- The camel of the Sahara Desert probably has more characteristics to enable it to survive than any other desert animal.
    - They have long eyelashes and can close their nostrils in sandstorms.
    - They can go without water for months and drink a lot of water very quickly when they find it.
    - They store fat in their humps. When digested it releases water.

Exam tip

Know a case study of an area of hot desert.

Example food chains in the desert are:

sun	➔	**primary producer**	➔	**plant eater**	➔	**predator**
sun	➔	desert grass	➔	springbok	➔	cheetah
sun	➔	desert grass	➔	springbok	➔	lion

Because both lions and cheetah eat springbok, they would be linked on a food web.

Economic development and environmental risk in hot deserts

Economic activity	Examples	Environmental risk
Irrigated agriculture using water from exotic rivers or boreholes and small dams, e.g. at oases	Nile Valley, Sahara Desert Orange River, Hardap, Namibia	Lowering the water table, supplies may dry up. Conflicts over water use between tourism and agriculture.
Mining	Uranium and diamonds in the Namib Desert	Unsightly and dangerous large hollows. Uses very large amounts of water. Damage to delicate vegetation.
Tourism, including ecotourism, because of the sunny climate, dramatic landscapes and wildlife	Namib Desert, especially the giant sand dunes at Sosselvei	Off-road vehicles damage the vegetation. Tourists can over-use water.
Development of urban areas often based on tourism	Las Vegas, and Palm Springs, Mojave Desert	Destruction of the desert ecosystem.

Recap

Hot deserts lie between 15° and 30° of latitude on the west side of the continents, except for the Sahara which stretches across to the east coast of Africa. They are hot and dry all year with cloudless skies. Summer days are very hot, but nights are cold.

Plants and animals exist in these difficult conditions because they have special adaptations. Plants also have to contend with loose sandy soils, or rocky or stony soils that are difficult for roots to penetrate.

See www.oxfordsecondary.com/esg-for-caie-igcse for the answers to the 'Apply' task.

Apply

1. Explain why the daily range of temperature for Khartoum (15½°N), in the Sahara Desert, is 14 °C in June (daytime maximum of 41.3 °C and night time minimum of 27.3 °C).
2. Study the rainfall data, taken over a period of 34 years, for Aswan (24°N) in the Sahara Desert. Aswan is usually described as having no annual rainfall.

Mean annual rainfall (mm)	Maximum rainfall in a year (mm)	Minimum rainfall in a year (mm)	Maximum rainfall in 24 hours (mm)
0.22	27.2	0	8.6

Using the information given, describe the rainfall characteristics of Aswan.

3. Suggest difficulties that would have to be overcome to set up a mine in a hot desert.

Review

- Hot deserts are dominated by high pressure in which air sinks and warms. There can be no condensation in sinking air.
- The wind then blows out as offshore trade winds, which are not able to pick up moisture over land, so the deserts are dry all year.
- Cold currents offshore increase aridity.
- Occasional rain falls in torrential convectional downpours.
- The constantly high angle of the sun causes the climate to be hot all year.
- Cloudless skies cause very high daily temperature ranges.

Raise your grade

Sample question

1. For a tropical desert you have studied, explain the factors that have influenced its climate. **[7 marks]**

Analysis

✓ To answer this question, you need to think about different types of factors that influence temperatures, rainfall and other climatic elements.

✓ Try to include a range of factors, at least three, in the answer and make sure you link the climate to a factor in every sentence you write.

Student answer

Hot deserts are found on the west sides of continents in latitudes around 30° from the Equator, where the air is sinking in high pressure systems. As the winds blow out over land from these, the air does not pick up any moisture, so deserts have little rainfall. They are hot because the full power of the sun reaches the ground, as skies are cloudless. [Level 2: 4 out of 7]

See www.oxfordsecondary.com/esg-for-caie-igcse for the mark scheme for this question.

Examiner feedback

The answer fails to state which tropical desert is the subject of the answer. There is one well developed link between rainfall and winds, where the answer explains one reason why the winds do not have the moisture needed for rain. That sentence lifts the answer into the low end of Level 2.

The first sentence does not attempt to answer the question, as it mentions factors but does not link them to the climate.

The last sentence does not mention factors that lead to absence of cloud so, apart from one sentence, these are simple undeveloped statements. The fact that cloudless skies have nothing to reflect, scatter or absorb solar radiation should be stated.

Economic development

Chapters 20 to 27 are designed to help you know and understand how and why countries or regions are at different levels of economic and industrial development and how our food, energy, and water supplies are produced. Problems associated with economic activity, such as soil erosion, climate change, desertification and different types of pollution are also included.

There will be two questions on *Economic development* in the *Cambridge IGCSE® or O Level* Paper 1 and there will be at least one question in Paper 2. The section on the risks of economic development has important links with the sections on agriculture, industry and tourism. These links may be emphasized in the examination questions.

Key ideas

- The lives of people in different countries are not equal.
- There are different ways of measuring inequality, including the Human Development Index (HDI).
- The types of jobs that people do changes as a country becomes more developed.
- Globalization has affected people's lives through wealth and employment, culture, communication and migration.
- The giant companies, known as transnational corporations (TNCs), have become increasingly important.

Countries at different stages of development

	More economically developed country (MEDC)	Less economically developed country (LEDC)
HDI	High	Low
GDP per capital ($US)	High	Low
Death rate per 1000	Low	High
Infant mortality per 1000 births	Low	High
Birth rate per 1000	Low	High
Life expectancy at birth	High	Low
Population growth rate (%)	Low	High
Adult literacy (% of population)	High	Low
Doctors per 1000	High	Low
Urban population (% of total)	High	Low
Agricultural employment (% of population)	Low	High
Assess to electricity (% of population)	High	Low
Internet use (% of population)	High	Low

Exam tip

You should be able to explain how the following factors affect a country's development: location, size, natural hazards, climate and soil, stable government, economic policies, ability to trade, and population issues.

Common error

Remember that inequalities exist not only between countries, but also between regions within countries.

Industrial sectors and employment structure

There are four main industrial sectors: **primary**, **secondary**, **tertiary** and **quaternary**. By analysing these sectors in any country or region, we can learn about its **employment structure**.

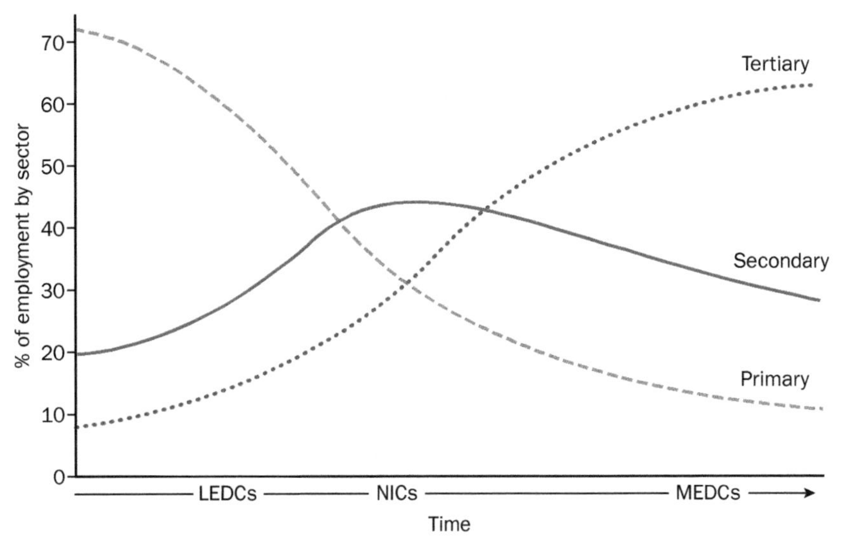

Fig. 20.1 *How employment in industrial sectors changes with time as a country becomes more developed*

Exam tip

Remember that employment statistics do not always recognize the quaternary sector and quaternary jobs are sometimes included in the secondary or tertiary sectors.

Exam tip

Learn how employment structure changes as a country becomes more developed.

Exam tip

Questions on employment structure sometimes use triangular graphs. Make sure that you know how to plot and read this type of graph.

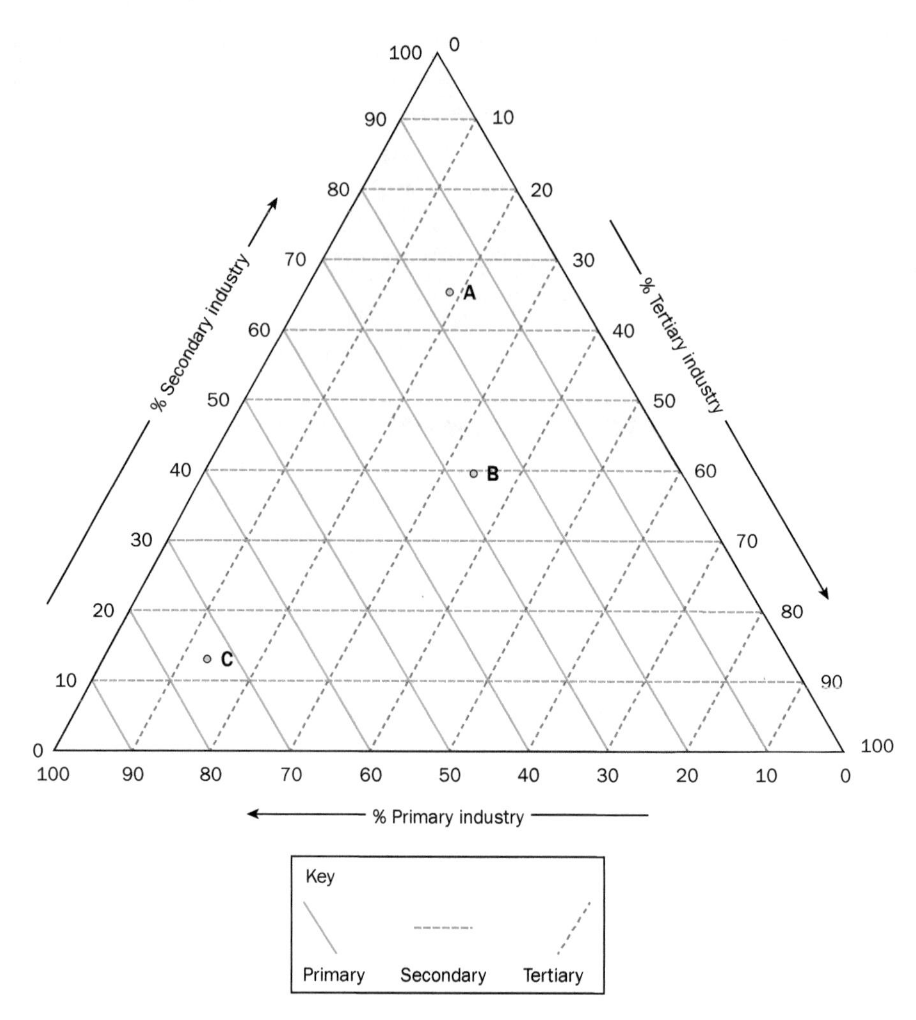

Fig. 20.2 *The employment structure of three countries. Which of the three countries is the least economically developed?*

Globalization

- The increase in world trade in goods and services is a feature of **globalization**.
- Countries are more affected by economic change in other countries.
- Cultures in different countries are becoming more similar in languages, food, music and clothing.
- There has been a change in the location of some manufacturing industries from MEDCs to LEDCs and NICs.
- There are worldwide environmental effects, such as air pollution and global warming.
- International population migration has increased.
- There has been a growth in transnational corporations (TNCs).
- Some of the world's great cities, such as London and New York, have become important beyond the boundaries of its own country. The transnational corporations have their headquarters in these cities, from where they control their businesses around the world.
- There have been advances in transport, notably in air travel.
- The containerization of freight is a sign of advancing globalization.

- International organizations, such as the European Union, United Nations and Commonwealth of Independent States, involve co-operation between countries in economic and military activities.
- Advances in communications infrastructure, such as the internet and cell phones, have allowed the rapid movement of knowledge and information.

Transnational corporations (TNCs)

Advantages of TNCs for LEDC host countries

- The presence of a transnational corporation provides jobs for local people.
- It provides a guaranteed income for people.
- It improves people's skills.
- It brings in foreign currency, which helps the country to develop.
- The increased employment also increases the demand for consumer goods in the LEDC and helps other industries to develop there.
- It can lead to the development of local raw materials, such as mining minerals or growing crops.
- It often leads to the development of infrastructure projects, such as roads, dams, airports, schools and hospitals.

Exam tip

Make sure that you know a case study of a TNC, including its organization and its impact on the host country.

Disadvantages of TNCs for LEDC host countries

- Most of the profits go abroad and are not reinvested in the host country.
- The numbers of local people employed can be small.
- The transnational corporation might suddenly decide to leave the LEDC, if conditions inside or outside the country change. This decision is made outside the LEDC.
- Raw materials, such as minerals, are often exported and not processed in the LEDC.
- Levels of pay are lower than elsewhere in the world.
- The operations of the company may cause environmental damage.

Impacts of TNCs in MEDCs

- Areas involved in manufacturing industries have suffered when TNCs have moved production to places with cheaper labour, often in LEDCs. This has led to unemployment and the economic decline of some regions.
- TNCs have often located their headquarters in 'world cities', where global brands are managed. This has increased skilled employment in management, accountancy, legal services, marketing and IT. Economic growth has occurred in these cities.

Recap

- Development can be measured by wealth and quality of life.
- There are inequalities within countries and between countries.
- As a country becomes more developed, the employment structure changes, with greater amounts of, at first secondary, then tertiary and finally quaternary employment.
- The world is becoming more interconnected. This is called globalization and has led to changes in people's wealth, employment, culture, communication and migration.
- The giant TNCs have become increasingly important in globalization.

Apply

Choose a TNC that you have studied. Make a list of the features of its organization and how it operates. Make another list of its impacts on the countries where it operates.

Review

These differences might be between countries or between regions of the same country.

Central 'core' areas	Regional 'peripheral' areas
More urbanized	More rural
More tertiary and quaternary industry	More primary industry
Higher incomes and more wealth	Lower wages and higher unemployment
Higher living costs	Lower living costs
Inward population migration	Outward population migration, especially of young educated workers – a 'brain drain'
Strong transport systems	Poor accessibility
Home of government and social elite	

Raise your grade

Sample question

1. The following graphs show the growth in world trade between 2005 and 2015.

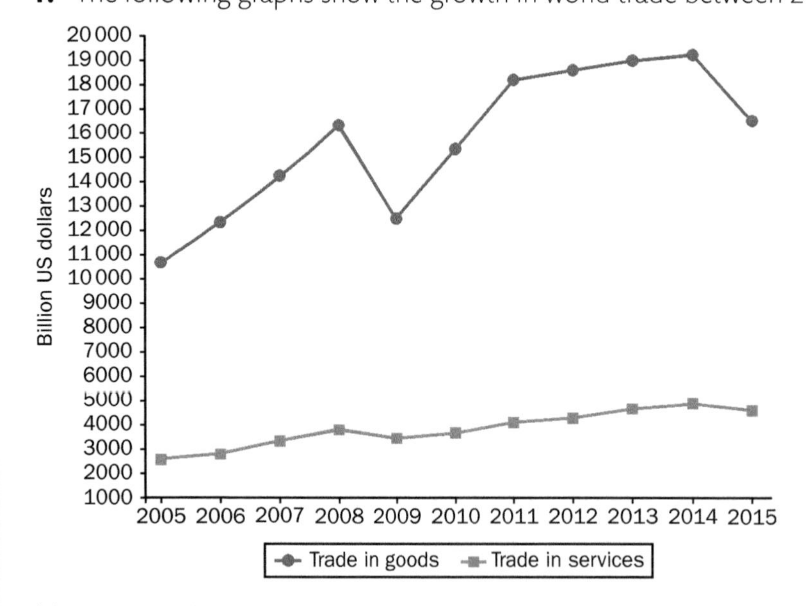

(a) How much was world trade in services worth in 2007? **[1 mark]**

(b) What is meant by the term *services*? Give an example. **[2 marks]**

(c) Describe the changes in world trade of goods and services. **[3 marks]**

Analysis

✓ Simple graph reading questions are often set in Paper 2. Be as accurate as you can and remember to give the correct units.

✓ The answer to part (c) could be written in two separate parts, one on goods and one on services. Quoting figures is generally useful. Answers should describe general trends and not every year.

Student answer

(a) 3100 [0 marks]

(b) Things that people use, like banking and finance. Physical goods are excluded. [2 marks]

(c) The trade in goods is worth more than the trade in services. The highest for goods is about US$18 billion and the highest in services is about US$5 billion. Goods had a steep drop in 2008 and services only have a slight drop. Goods drop in 2015, but services only change slightly. [1 mark]

Total: 3 out of 6

Examiner feedback

In part (a) the figure is correct but no units have been given, so no marks can be awarded. The answer to part (b) is very good. The answer to part (c) is about *differences*, but it should focus on *changes*. There is no need to compare the two. One mark has been scored for the decrease in 2008.

See www.oxfordsecondary.com/esg-for-caie-igcse for the mark scheme for this question.

Sample question

2. The following diagram shows how the relative size of industrial sectors has changed.

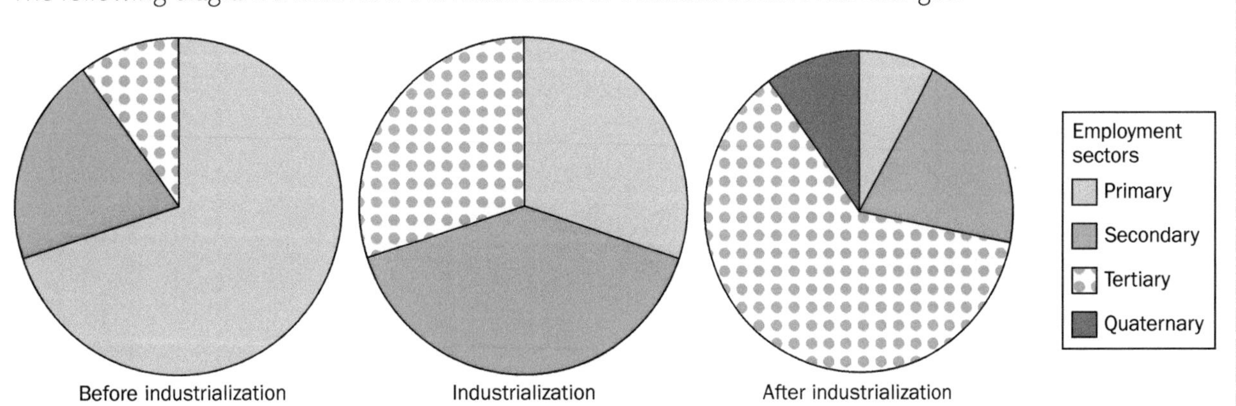

(a) Define the term *quaternary industry*. **[1 mark]**

(b) Describe the changes in the size of secondary industry during and after industrialization. **[2 marks]**

(c) Suggest *reasons* for the changes you have described in part (b). **[4 marks]**

Analysis

✓ Parts (b) and (c) are good examples of how the command words 'describe' and 'suggest' produce different answers.

Student answer

See www.oxfordsecondary.com/esg-for-caie-igcse for the mark scheme for this question.

(a) High-tech industries, like aerospace. [1 mark]

(b) Before industrialization, secondary industry accounts for a low percentage of employment – about 20%. During industrialization, this increases to become the largest sector – about 40%. After industrialization, it drops again to about 15%. [2 marks]

(c) As industrialization takes place, people move from rural areas to urban areas for jobs. The newly developed industries need a workforce. Mechanization of farming removes jobs in the primary sector. After industrialization, jobs in industry start to be replaced by mechanization and labour-intensive jobs are moved to LEDCs. The growing tertiary sector provides new jobs. [4 marks]

Total: 7 out of 7

Examiner feedback

This is an excellent answer. In part (c) the candidate makes more than the four necessary points. Compare the answer with the mark scheme.

1 Food production

Key ideas

- **Agriculture is a system with inputs, processes and outputs.**
- **Agricultural systems may be:**
  - **arable, pastoral or mixed**
  - **intensive or extensive**
  - **commercial or subsistence.**
- **Food shortages are caused by physical and human factors.**

Inputs, outputs and processes

Physical inputs (provided by nature)

- Climate: temperatures, rainfall, sunshine
- Soil
- Land and its relief

Human inputs (provided by people)

- Capital
- Labour
- Machinery and tools
- Seeds
- Social structures
- Government influence
- Market influence
- Fertilizers, pesticides and herbicides
- Irrigation

Outputs

- Crops
- Meat
- Milk
- Industrial products such as cotton, rubber or leather

Processes

- Preparation of the land: clearing vegetation, providing terracing, drainage and irrigation systems
- Ploughing
- Sowing
- Weeding
- Application of fertilizers, pesticides, herbicides and irrigation
- Harvesting
- Storage and transporting to market

Exam tip

You should learn a case study of a farm or an agricultural system. This could be commercial or subsistence, intensive or extensive.

Candidates often confuse the physical inputs with the human inputs.

Commercial farming and subsistence farming

	Commercial farming	Subsistence farming
Capital (money)	Large capital input, sometimes from international companies	A complete lack of capital may prevent any increase in output
Land	Large area	Very small farms
Labour	Paid labour (often skilled), much use of research and development	Family labour, relying on traditional methods
Machinery and tools	Much use of mechanization for all processes	Hand tools, such as hoes and ploughs, sometimes pulled by animals
Seeds	Improved varieties and hybrids	Seeds left over from the previous year's crop
Market influence	Production is geared to current market demands and prices	No market influence
Fertilizers	Generally used	Used much less, although sometimes animal manure is available
Pesticides and herbicides	Generally used	Used much less
Irrigation in dry areas	Uses complex systems	Either none or very low-technology systems

Common error

Candidates often believe that commercial farming is always intensive. This is not true; commercial farming can be extensive or intensive, as can subsistence farming.

Shifting cultivation

Today this system is only practised in a few areas. An area of land is cleared, and ash from burnt vegetation is used as a fertilizer. The land is cultivated for a few years in the traditional manner until it is exhausted and crop yields decline. The people then move to another area, sometimes building a new settlement, and repeat the process, not returning to the original plot for perhaps 20 years.

This system is still used in some tropical areas, where the soil fertility is low and minerals are leached by heavy rainfall.

Intensive and extensive farming

	An intensive farm	An extensive farm
Area of land	Small	Large
Large machines	Few	Many
Labour input per hectare	High	Low
Fertilizer input per hectare	High	Low
Output per hectare	High	Low

Food shortages

Causes of food shortages

Food shortages are linked to crop failure and poor agricultural yields.

Physical causes	Human causes
Soil exhaustion as a result of overcropping, monoculture, insufficient fertilizer and manure	**Diseases affecting** farmers, such as malaria, HIV/AIDS
Drought, particularly in areas of the tropics where seasonal rainfall occurs	**Low capital investment** – many people who practise subsistence agriculture are stuck in a vicious circle of poverty
Floods – where farming occurs on floodplains, serious flooding can lead to the complete loss of a year's harvest	**Poor transport** – farmers in remote areas find it difficult to sell surpluses and raise capital, receive supplies and information about possible improvements to farming
Tropical cyclones – crops can be destroyed by strong winds, torrential rain or the associated floods	**Wars** – people are forced to leave their homes and become refugees, which has an effect on their ability to make long-term investments in increased food production
Pests include locusts and birds, which can destroy mature crops	**Increased use of biofuels** – from 2008 to 2011 some land previously used for the production of food was changed to produce crops for biofuel production
Crop diseases can destroy crops in the fields or during storage	
Animal diseases, such as foot and mouth disease, nagana or trypanosomiasis, result in low production of meat and milk or death of animals	

Exam tip

You should learn a case study of a country or region suffering from food shortages.

Common error

When naming the location of a case study, be as specific as you can. For example, it might not be accurate to name a whole country as an example of food shortages, but a region of that country would be fine.

Solutions to agricultural problems and ways of increasing the food supply

- New hybrid seed varieties can be very responsive to fertilizers, give higher yields and have shorter growing seasons (although these may bring other problems). Genetically modified (GM) crops can have the same effects.
- Extend irrigation in dry areas.
- Different crops, for example alternatives to maize include sorghum, sweet potatoes, cassava and groundnuts.
- Subsidized farming inputs, such as tractors, seeds and fertilizer, help overcome the lack of capital.
- Disease control, for example methods to control foot and mouth disease such as fencing.
- Education and training of farmers in new methods or growing different crops.
- Improved markets for crops to stimulate production.
- Measures to control soil erosion, including inter-cropping, terracing, contour ploughing, crop rotation and reducing stocking densities.

Recap

- Agriculture is a system with physical inputs (e.g. climate, soil, land and relief), human inputs (e.g. capital, labour, machinery and fertilizers) and processes (e.g. drainage, irrigation, ploughing and harvesting).
- Agriculture may be: arable, pastoral or mixed; intensive or extensive; commercial or subsistence.
- Food shortages are caused by physical factors (e.g. soil exhaustion, drought, floods, tropical cyclones, pests including locusts and birds) and human factors (e.g. diseases affecting farmers, low capital investment, poor transport, wars, increased use of biofuels).

Apply

Name an example of a farming system. For the example you have chosen, make a list of the:

- physical inputs
- human inputs
- processes
- outputs.

See www.oxfordsecondary.com/esg-for-caie-igcse for the answers to the 'Apply' task.

Review

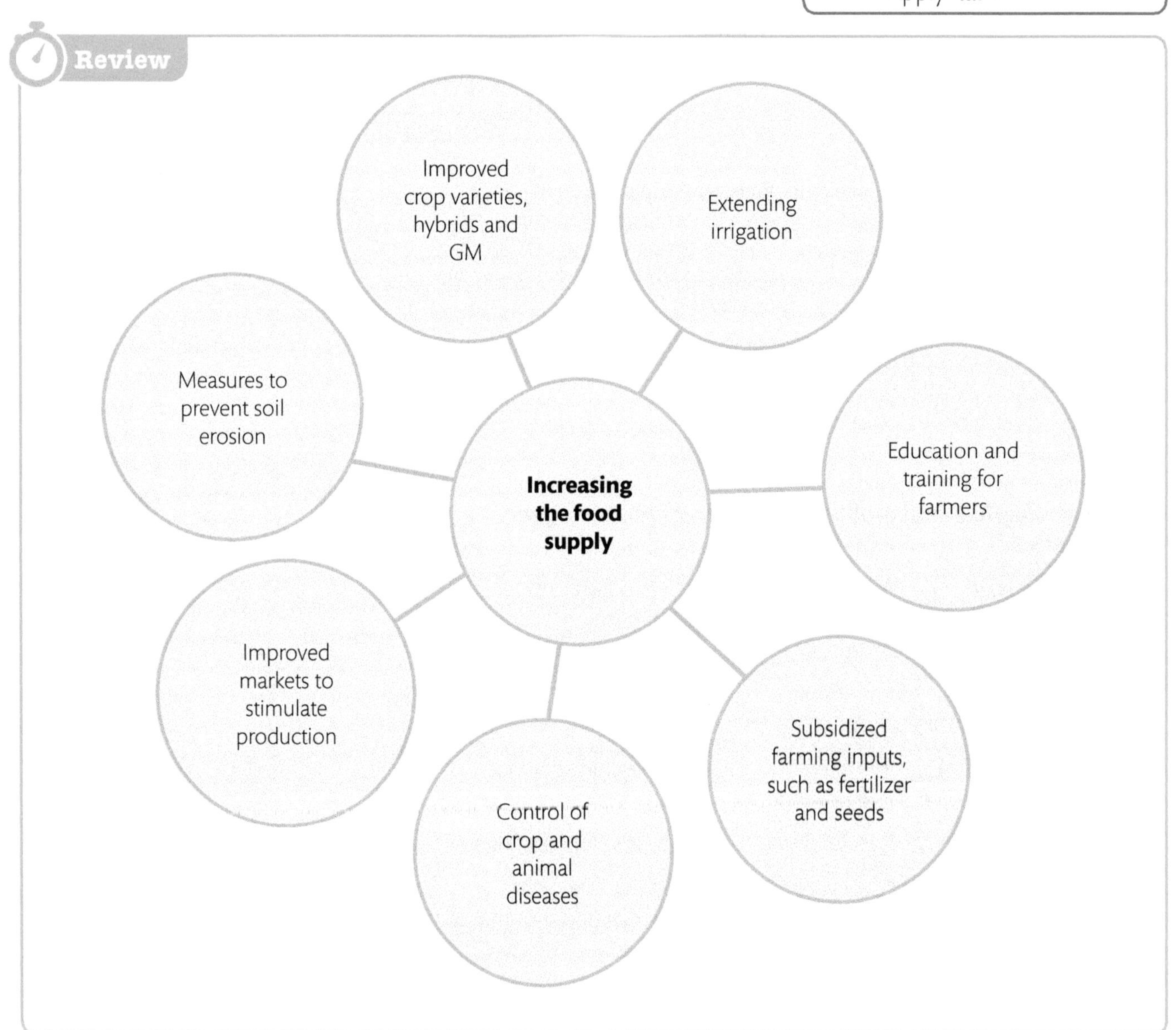

Raise your grade

Sample question

1. The table below gives information about arable farming in three countries.

Inputs	Borneo	Japan	California
Direct energy			
Labour	0.626	0.804	0.008
Axe and hoe	0.016		
Machinery		0.189	0.360
Vehicle fuel		0.910	3.921
Indirect energy			
Fertilizers		2.313	4.317
Seeds	0.392	0.813	1.140
Irrigation		0.910	1.299
Pesticides		1.047	1.490
Electricity		0.007	0.380
Transport		0.051	0.121
Total inputs			
Output-rice yield	**7.318**	**17.598**	**22.3698**
Energy efficiency ratio	**7.08**	**2.49**	**1.57**

(a) What is meant by the term *arable farming*? **[1 mark]**

(b) Using the table, describe the differences in rice yield (output) per hectare in Borneo, Japan and California. **[2 marks]**

(c) Using the table, suggest reasons for the differences in rice yield between Japan and California. **[3 marks]**

Analysis

✓ You must use your knowledge to answer part (a), but the answers to parts (b) and (c) are based entirely on information in the table.

Student answer

(a) Producing crops not animals. [1 mark]

(b) In Borneo, yields are low. In Japan, yields are more than twice this. California has the highest yields, more than three times those of Borneo. [2 marks]

(c) California has more use of machinery and fuel for the machinery. More money is spent on fertilizer, seeds, pesticides, irrigation and transport. [3 marks]

Total: 6 out of 6

See www.oxfordsecondary.com/esg-for-caie-igcse for the mark scheme for this question.

Examiner feedback

Although this is a short answer, all the relevant points are made.

Sample question

2. The diagram shows how subsistence farmers can be trapped in a cycle of low production.

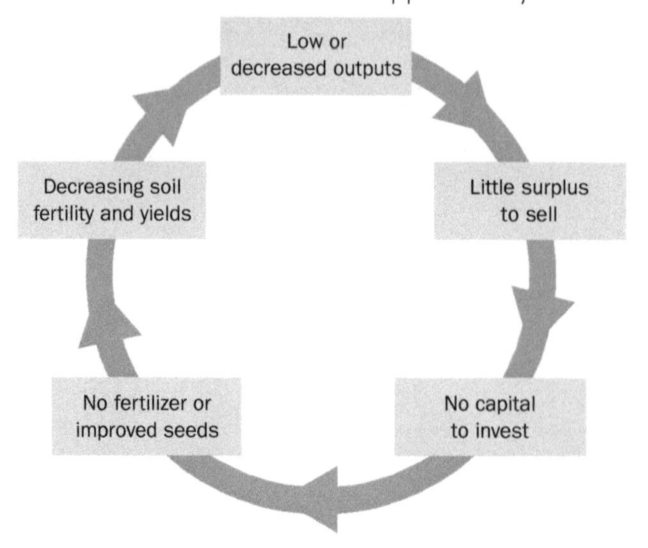

(a) What is meant by *subsistence farming*? **[1 mark]**

(b) Identify **two** inputs in a subsistence farm. **[2 marks]**

(c) Suggest how the cycle of low production might be broken. **[4 marks]**

Analysis

✓ This question mainly tests your knowledge of subsistence farming. You should know the inputs, processes and outputs of the system and the vicious circle of low production that subsistence farmers find themselves in.

Student answer

See www.oxfordsecondary.com/esg-for-caie-igcse for the mark scheme for this question.

(a) Farming to eat not to sell. [1 mark]

(b) Simple machinery and no capital. [1 mark]

(c) Providing farmers with food aid to feed their families would solve the problem. [0 marks]

Total: 2 out of 7

Examiner feedback

The definition in part (a) is correct. In part (b) simple machinery gains one mark, but saying 'no capital' does not gain a mark. Avoid negative comments like this.

Part (c) is not correct as it would not break the cycle. The correct points are listed in the online mark scheme.

2 Industry

Key ideas

- **Industrial systems involve inputs, processes and outputs.**
- **The location of factories is influenced by a variety of physical and human factors.**
- **The scale of production and the way it is organized is influenced by a variety of physical and human factors.**

Types of industry

Manufacturing industry

- In a **manufacturing** industry, products are usually made in large quantities in a factory using machinery.
- Domestic industry is small-scale production in the home, mainly using hand tools.

Processing industry

Processing is done for a variety of reasons:

- to deliver for direct sale to consumers in a protective container, such as the canning of fruit.
- for direct sale to consumers as a safer product, such as the pasteurization of milk.
- to change a perishable natural product into a longer-lasting form, such as making cheese from milk and the tanning process which converts skin into leather, which will not quickly decay.
- to turn it into a form which can be further changed in a manufacturing industry, such as leather into shoes.
- to reduce transport costs of raw materials. Some waste is usually removed from mineral ores near mines to reduce bulk, so that it is cheaper to transport for manufacturing.

Assembly industry

An example of the **assembly** industry is the aircraft industry. Planes are made from parts made elsewhere and put together on an assembly line.

High-technology industry

- **High-technology** is characterized by a highly automated and computerized manufacturing process.
- It uses a high degree of research and development to develop new products and design new machines to make them, as it is desirable to keep ahead of competitors.
- High-tech products include pharmaceuticals, precision instruments, computers, televisions, mobile phones and aircraft.
- Biotechnology companies develop new kinds of food, drink and vaccines.

Common error

The products of high-tech industry are not always high-tech themselves. For example, pharmaceutical products are low-tech, but produced using high-tech methods.

Industrial systems

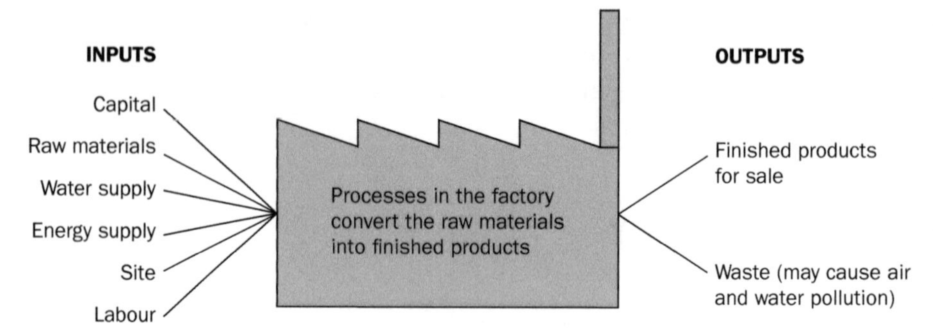

Fig. 22.1 *The industrial system of inputs, processes and outputs*

The influence of raw material inputs on the processes and outputs of industrial systems

Some industries have raw materials that require a lot of processing, which can lead to a lot of pollution.

Industry	Raw materials	Processes	Outputs	Possible adverse effects
Iron and steel manufacturing	Iron ore, coke, limestone to separate the iron from impurities in the ore, water, recycled scrap iron. For special steel: alloys such as chromium and cobalt	Heating of ore to separate the iron by burning coke, further heating and processing to make steel, rolling and cutting	1. Cast iron, pig iron and steel 2. Waste: slag and gases (sulfur dioxide, carbon dioxide, nitrous and nitric oxides, hydrogen sulfide)	Noise, large ugly buildings, slag heaps, dust, air pollution, water pollution (contaminated cooling water and scrubber effluent), risk of fires and explosions

Factors affecting industrial location

Physical factors

Raw materials	• Raw materials for heavy industries are bulky and expensive to transport. Nearness to iron ore or coal mines or to ports, if imported, used to be the most important factor in their location. • Today's transport systems are more economical and bulky commodities are transported across the globe in large ships. • Market and government influence have become more important now in the locations of vehicle manufacture. • Light industries use materials and have products which are small in volume but of very high value, so transport costs are of little importance.
Site	• This is important for large-scale manufacturing, such as iron and steel making, oil refining or chemical manufacture. • The large size of the plant means that very large areas of flat land are needed. • Sites need to be well-drained and on solid bedrock.
Energy	• Energy supplies were important in the past when sites next to fast-flowing rivers or coal mines were favoured. Today, a link to the electricity grid is often sufficient. • For a few manufacturing industries, e.g. aluminium smelting, which require very large amounts of power, access to cheap energy is important.
Water supply	• Some industries, such as paper, chemicals and metals, require water in high quantities and may need to be located where they can have their own supplies from rivers or boreholes.
Natural harbours and route centres	• Ports are favoured locations because raw materials can be imported and products exported at less cost. • Major roads and railways often follow natural routes such as valleys. Industries locate along them and at junctions of routes.

Human and economic factors

Capital	• The start-up costs of an industry may come from other businesses, banks or governments. Finance may be more freely available in some countries or areas than in others, and often connected with political factors. • Only global transnational companies have access to the capital needed for the motor vehicle industry, which achieves economies of scale by mass production on very large assembly lines using complex machinery.
Labour	• The size of the labour force is important in many industries. Usually, the quality, adaptability and reputation of the labour force is just as important as its size. • Large numbers of skilled workers are needed for vehicle manufacturing to operate highly complex production lines, so the company seeks educated, hard-working employees in countries with few restrictive employment laws. • Workers in LEDCs are about 40% cheaper than in MEDCs.
Transport	• Transport is still an important factor in the location of industries producing bulky goods. • For vehicle manufacturing, access to ports and good road systems is important for the assembly of components and distribution of vehicles to markets. • The massive increase in the use of containers, which can carry many items securely and cheaply, has helped the development of industries in Asia's NICs. • TNCs are able to take advantage of the low production costs (cheaper wages) there and to use containers to transport the products to the rich markets in MEDCs.
Markets	• This is the most important factor in the location of the motor vehicle industry today; there has been rapid expansion into NICs with very large populations, such as China, India, Brazil.
Political influence	• Governments provide financial incentives to companies to locate in particular areas. • The tax system of a country influences decisions taken by transnational companies. • As many countries have higher taxation levels on imported manufactured goods than on components, it is cheaper to transport the components for motor vehicles and assemble them in a country where there is a large market, than to export cars to the country.
Quality of life	• This influences the location of industries which require a highly-skilled professional workforce that prefers pleasant areas with good housing and leisure facilities.

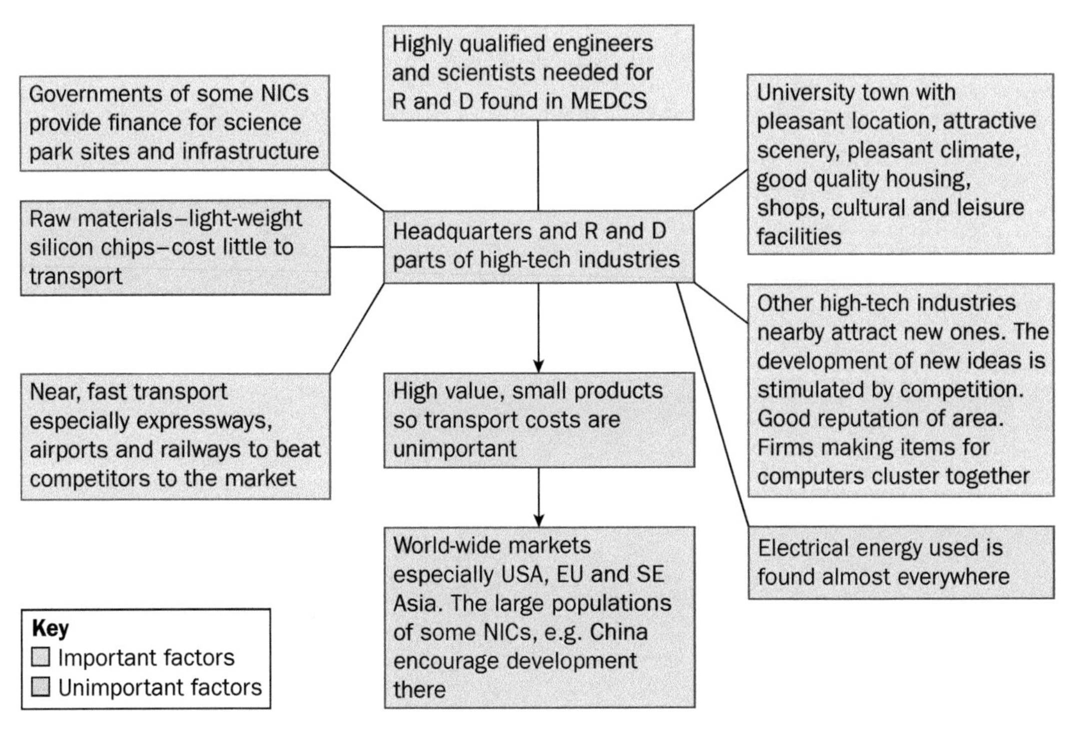

Fig. 22.2 *The relative importance of inputs to the location of high-tech industry*

Exam tip

Know about an industrial zone or factory. Be able to name specific facts about its type of industry, factors that influenced its site, production, inputs, processes and outputs. Be able to name some specific factors, such as roads, ports, rivers or other water sources, energy supply and source of the labour force.

Industrial zones

Industrial zones with many manufacturing industries

Industrial zones developed where there were very favourable circumstances for industrial development:

- large areas of flat or gently sloping land
- an important mineral, usually iron ore, in the area with a local energy source, such as coal or oil.

These were the basis for the development of many manufacturing industries, such as engineering and chemical industries. Many industries use the products of oil refining, chemical, and iron and steel industries as their raw materials, so it is cheaper to locate near them. People move into the area, attracted by jobs in the industries, so a market develops for their products. Industries favoured a local port or railway junction, as bulky raw materials are moved more cheaply by rail than road and even more cheaply by ship.

Examples include the Ruhr area of Germany, the area around Tokyo in Japan, and the Pittsburgh area in the US.

Industrial zones based on high-tech industry

These more recent industrial agglomerations have a different basis. Jinan in NE China is one example of many. Its advantages for development were:

- a large, flat site
- the government initiative that established it
- government-funded improvements to the transport infrastructure with expressways that link to the nearby superhighway between Beijing and Shanghai and to Qingdao – one of the world's main container seaports
- a government-funded international airport and railway station no more than 20 km away
- a good telecommunications network
- 37 universities and colleges to provide a ready supply of educated workers
- schools, nurseries and other facilities that workers like to have nearby
- computer manufacturers that attracted a lot of software industries to set up close by
- an attractive mountain area on its outskirts and many trees in the city.

Recap

- Industries are classified into four types: manufacturing, processing, assembly and high-technology.
- Processes change inputs into outputs.
- Inputs include raw materials; capital; water and energy supplies; site requirements and labour. Government policy can provide land, advance factory units and capital.
- Outputs include the products and also waste, such as used water, smoke and other gases.
- Transport is not an input to the industry itself, but is required to bring in raw materials and labour, as well as to take products to market.

Apply

For a named industrial zone or factory you have studied, describe the industry type(s) and factors influencing its location. Include at least four specific details, such as source of raw materials, market, source of capital input and named routes.

See **www.oxfordsecondary.com/esg-for-caie-igcse** for the answers to the 'Apply' task.

Review

The headquarters of high-tech industries have different influences on their location from most industries. The most important are:

- an attractive university town for highly qualified scientists and engineers for R and D
- sites on science parks near fast transport to beat competitors to market
- quick access to an airport to attend conferences and meetings on their different sites, and visit branch factories and universities conducting research for them
- proximity to other high-tech industries to keep in touch with competition.

Different factors are more important for other industries, such as:

- large-scale manufacturing and industries with assembly lines need a large, flat site
- most processing industries locate near the raw material they process
- some industries have to locate near cheap power or reliable water supplies
- government influence, access to labour of the right type and access to transport and markets.

Sample question

1. Study the pie graph of the types of industry at the Oxford Science Park, UK. The Oxford Science Park has more than 70 companies with more than 2500 employees and is owned and managed by Magdalen College, Oxford University.

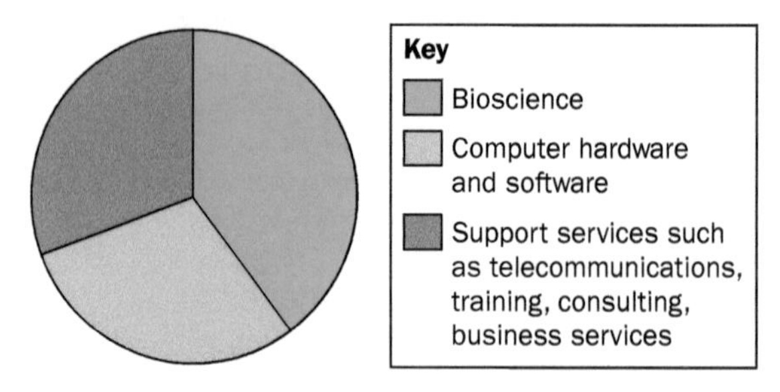

(a) What percentage of industries on the Oxford Science Park are bioscience industries? **[1 mark]**

(b) Explain why the types of industry shown on the graph locate close to major universities. **[2 marks]**

(c) Modern science parks are designed to attract high-tech industries and their employees. Suggest ways in which the Oxford Science Park is likely to do this. **[4 marks]**

Analysis

✓ The factors governing the location of high-tech industries are very different from those needed by heavy industries, such as iron and steel.

✓ You will be expected to apply your knowledge of a science or business park you have studied to this unfamiliar situation.

Student answer

(a) 40% [1 mark]

(b) Nearness to a university will enable them to employ graduates, because they need to have workers with innovative ideas to keep ahead of rivals. Research and development is of the greatest importance to high-tech companies. [1 mark]

(c) The science park will be planned to provide an attractive environment to the workers and all their requirements to give them an accessible and good working environment. The buildings will be positioned with car parks, gardens and lawns around them. There will be walkways away from roads and access to jogging paths. Showers will be provided on site, as well as a variety of bars and restaurants. The site will have been chosen to be easily accessible from airports by motorways or railways. There will be good quality hotels nearby. [3 marks]

Total: 5 out of 7

Examiner feedback

Part (a) is correct and part (b) scores one mark, but not a second one because it is all about the high quality of the workforce. Three detailed points are made for part (c), but the reference to transport and hotels is outside the park itself, so is not relevant to this question, even though they would be important considerations when the site for the park was selected.

Key ideas

- There has been an enormous growth in tourism in recent decades.
- Tourists travel to see attractions of physical and human landscapes.
- Tourism brings both benefits and disadvantages to tourist areas.
- Careful management is needed if tourism is to be sustainable.

Physical attractions

- Attractive scenery – sightseeing
- Physical features of special interest – geysers, potholes, caves, blowholes, volcanic craters, etc.
- Beaches in areas with warm, sunny climates
- The sea – swimming, surfing, fishing, yachting
- Wildlife – including safaris, whale watching
- Countryside suitable for activity holidays: mountaineering, fell walking, kayaking, white-water rafting, horse riding, skiing and snow boarding

Attractions of the human landscape

- Historical buildings and artefacts
- Well-known landmarks
- Sites of famous historical events
- Different cultures
- City breaks to see historical settlements, shops, museums, art galleries, etc.
- Gardens open to the public
- Organized events, e.g. film festivals, international sports fixtures and tournaments
- Theme parks, water parks and zoos

Advantages and disadvantages of tourism

Economic benefits of tourism	Negative economic effects of tourism
There is reduced unemployment. Standards of living rise with greater employment and income.	There is seasonal unemployment, e.g. in the hurricane season of the Caribbean.
Businesses connected with tourism, such as hotels and coach companies, make a profit and employ more staff. Local craft industries benefit as tourists buy souvenirs. Farmers sell food to hotels.	Many hotels and other businesses connected with tourism are owned by foreign companies, so much of the profits go to other countries.
Other local businesses, such as shops, benefit from the increased wealth and spending power in the area. They take on more workers.	People move to the coasts to work in tourism, so the inland economy, e.g. farming, declines.
Employees and businesses pay taxes and increase the wealth of the country, so more development of the country's infrastructure and services can be funded.	Foreign workers often send their wages home.
Skills learnt while working in tourism can be used in other occupations and areas of the country.	Skills gained through jobs in tourism enable some to leave the country to work in other countries.
Tourist areas become richer and able to provide more facilities, including health and education. (Tourism provides nearly 60% of the GDP of the Maldives.)	Parts of the country without tourist attractions remain poorer. Inequalities increase between the richer tourist resorts and other areas.
Foreign exchange increases as tourists change money into the local currency. They pay a fee to do so and the exchange rate is always favourable to the country.	In periods of economic downturn in MEDCs, businesses and workers in tourism lose income because fewer tourists travel.

Common error

It is important to remember that tourism brings advantages as well as disadvantages for the local people and economy.

It is important to be specific about why tourism is beneficial to the economy.

Exam tip

Know a case study of an area where tourism is important. Be able to give examples of its physical and human attractions and to describe and explain the benefits and disadvantages of tourism to the area.

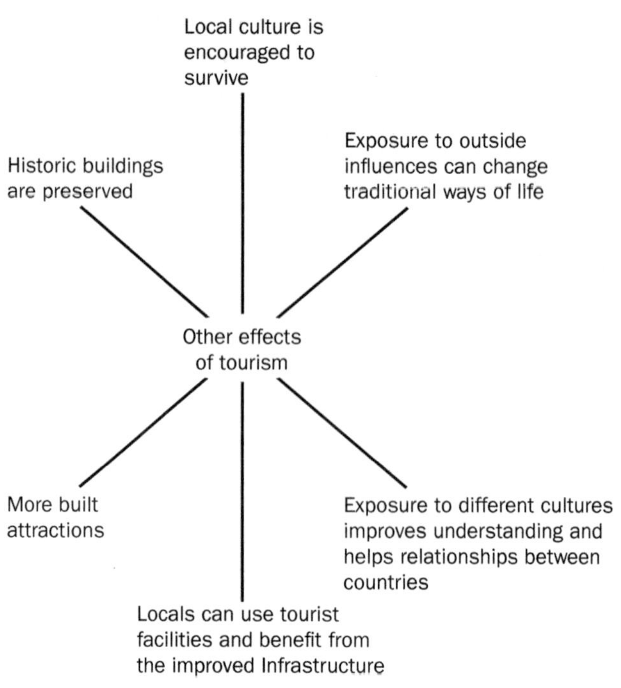

Fig. 23.1 *Other effects of tourism*

Causes of declining tourism

Tourist numbers to a country can reduce suddenly due to several reasons.

- Political instability. In recent years, the economy of some countries has declined markedly as civil unrest and war have deterred tourists from visiting. Tourism used to provide 80% of Egypt's GDP. It now fluctuates around 10%. Revenues from tourism fell from a record high in 2010 to a record low in 2016, costing the government the equivalent of more than US $8.5 billion.
- Natural hazard events. The series of hurricanes that devastated Caribbean Islands like St Maarten, Anguilla and the British Virgin Islands in 2017 will reduce visitor numbers for several years.
- Disease outbreaks. In 2016, pregnant women were advised not to travel to parts of South America that were affected by an outbreak of Zika virus, which causes head and brain damage in foetuses and babies.
- Decisions by local authorities to reduce numbers. Residents of cities such as Barcelona, Amsterdam and Venice want to stop **mass tourism** disrupting their lives. Measures include restrictions on the numbers of apartment rentals, hotels and tourist shops, turnstiles on crowded streets, increasing the tourist tax and banning noisy activities. Venice is also proposing to ban large cruise ships from docking near the centre because of damage done to the banks of the waterways.

Many LEDCs rely heavily on the income from tourism, but some are trying to reduce the problems by promoting small-scale and sustainable **ecotourism** and **community-based tourism**.

Recap

- World total tourist arrivals nearly doubled from 681 000 in 2000 to over 1 200 000 in 2017, partly because of improved air travel, the increase in numbers of pensioners and more high earners in rapidly developing countries (like China and South Korea).
- Tourists are attracted by physical features, such as a warm, sunny climate and also by human attractions (such as Disneyworld).
- Tourism brings large amounts of revenue to MEDCs, but is a small part of their total income. Whereas smaller revenue is earned by LEDCs but forms a much larger proportion of their total income.

Apply

Prepare for a question about your chosen case study of an area where tourism is important. Consider what its attractions are and the benefits and disadvantages of tourism for the area. Include specific detail where possible.

See www.oxfordsecondary.com/esg-for-caie-igcse for the answers to the 'Apply' task.

Review

- Mass tourism has disadvantages and many countries are realizing the need for sustainable tourism by limiting numbers visiting hotspots, closing off eroded areas (like sand dunes to allow vegetation to regenerate), restricting tourist access to certain paths, restricting building, and encouraging ecotourism and community-based tourism.
- Mass tourism is not allowed in the Galapagos Islands National Park, where strict measures are taken to protect the unique animal and plant ecosystem. Tourists can only travel on small boats, the routes of which are determined by the authorities and only four boat landings are allowed on each island each fortnight. Also, tourists walk on marked trails accompanied by local guides and are not allowed within 2 metres of wildlife.
- There are case studies of tourist areas and numerous photographs of attractions in Chapter 6 of *Complete Geography for Cambridge IGCSE and O Level*.

Sample question

1. Study the graph showing tourist arrivals on an island at about 40 °N.

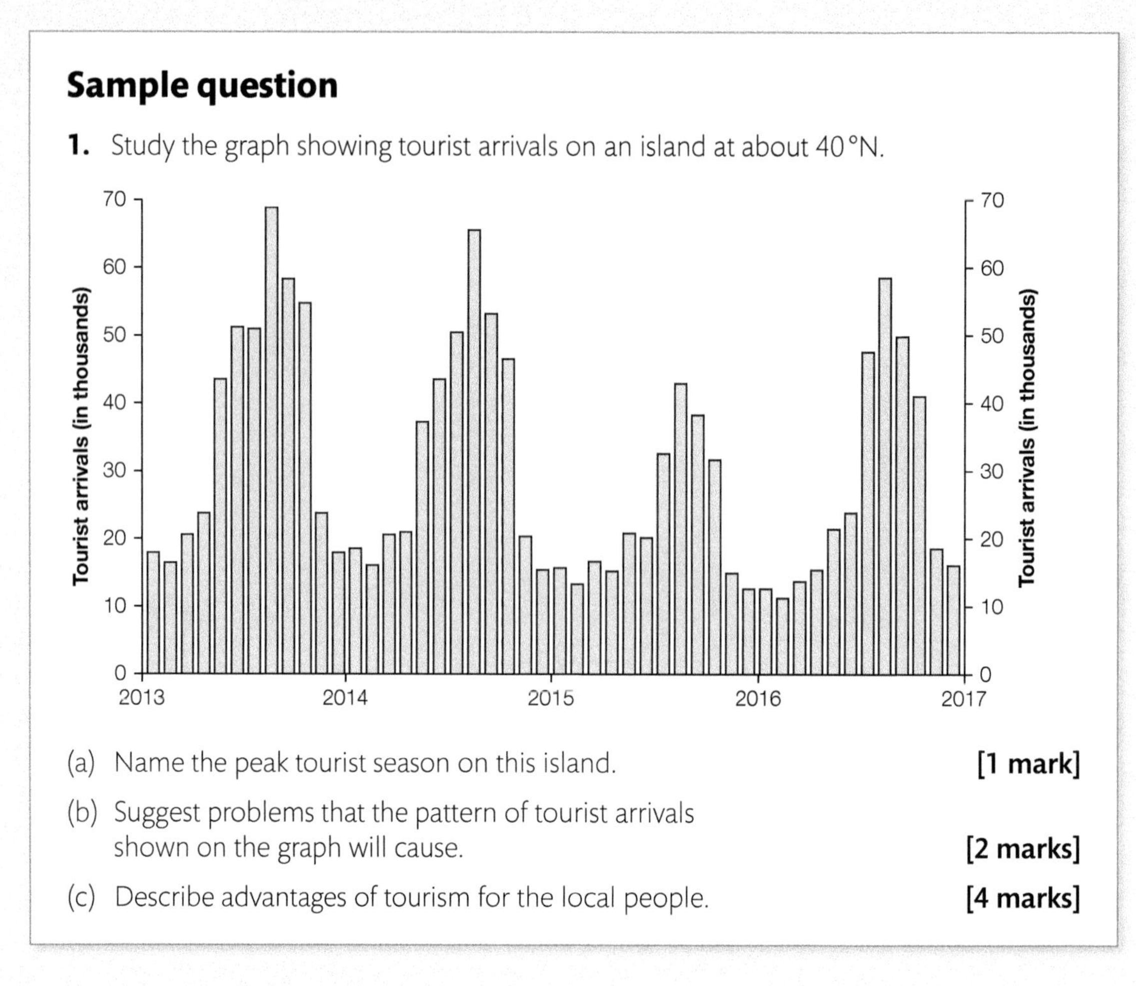

(a) Name the peak tourist season on this island. **[1 mark]**

(b) Suggest problems that the pattern of tourist arrivals shown on the graph will cause. **[2 marks]**

(c) Describe advantages of tourism for the local people. **[4 marks]**

See www.oxfordsecondary.com/esg-for-caie-igcse for the mark scheme for this question.

Analysis

✓ This question concentrates on some of the benefits and disadvantages of tourism.

✓ The pattern shown on the bar graph is typical of many tourist areas with seasonal differences in weather.

Student answer

(a) August [0 marks]

(b) Restaurants, hotels and bars will be very busy in the main summer season. There may be a shortage of workers to cope and people may have to be enticed in from other areas with high wages, raising prices for locals as well as tourists. Everywhere will be more congested and polluted. [2 marks]

(c) Tourism will bring prosperity. Secondly, it will enable the locals to learn about different cultures and so increase understanding of them. Thirdly, locals may learn from the knowledge their visitors bring. For instance, they may improve their language skills. Locals will also be able to use amenities provided for tourists, which can range from better bus services to cinemas, theatres, water parks and theme parks. The government will obtain revenue from tourism to boost the economy. [2 marks]

Total: 4 out of 7

Examiner feedback

In part (a), August is not a season; this error is not uncommon.

In part (b), maximum marks have been gained, but the more serious problem caused by up to 50000 fewer tourist arrivals in the winter months has been omitted. More serious is the downturn in numbers over a whole year in 2015. Hotels will have unoccupied rooms, even in the peak tourist season.

In part (c), the first advantage needs to be detailed - more jobs are available, for example driving taxis and tourist buses, acting as tour guides and representatives of holiday companies or working in restaurants and hotels, giving more income and spending power. Local businesses will benefit from the increased spending power of the locals employed, as well as the tourists.

Government revenues are not relevant to the question which is about benefits for the local people.

Sample question

2. Study the map, which shows some of the tourist attractions of Namibia, a country in Southern Africa.

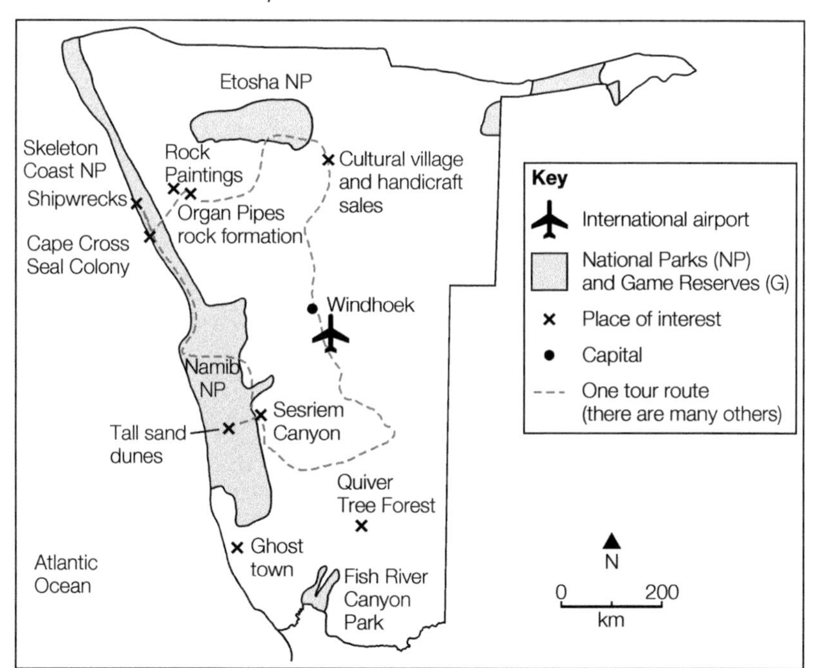

(a) Estimate the distance covered by the tour route shown on the map. Select the nearest to your answer from 1000, 1500, 2000 or 2500 km. **[1 mark]**

(b) Using map evidence, state **one** main physical attraction and **one** main attraction of the human landscape seen on the tour. **[2 marks]**

(c) Most of Namibia is tropical desert and semi-desert. Suggest **four** different types of problems for Namibia caused by tourism and **one** measure that could be taken to reduce each. **[4 marks]**

Analysis

- ✓ You need to be able to explain the growth of tourism in relation to the physical and human landscape.
- ✓ This question also tests your knowledge of the disadvantages of tourism to tourist areas and ways in which they could be managed.
- ✓ It is vital that you know the meaning of common geographical terms, such as 'physical'.

Student answer

See www.oxfordsecondary.com/esg-for-caie-igcse for the mark scheme for this question.

(a) Many kilometres [0 marks]

(b) Etosha National Park, rock paintings [2 marks]

(c) The noise tourists make could frighten the game animals so they should be kept at a distance from them. Vegetation on the sand dunes could be easily uprooted, so public access should be restricted to set routes. Tourists could offend local people by the way they dress so should be advised how to dress so as not to cause offence. The tour buses will cause air pollution. [3 marks]

Total: 5 out of 7

Examiner feedback

The answer to part (a) is not an answer to the question asked.

Part (b) is correctly answered.

There are three problems with solutions in part (c), but the air pollution point did not have a solution and the Namibian government would be reluctant to use one because tourism is vital to the economy.

Key ideas

- World energy consumption is increasing, especially in LEDCs.
- Energy consumption varies from country to country, depending on the level of economic development.
- Non-renewable energy supplies include coal, oil and natural gas.
- Renewable energy supplies include geothermal, wind, hydroelectric power, wave and tidal power, solar power and biofuels.
- Each different energy supply has its own benefits and disadvantages.

World energy consumption

Fig. 24.1 *Energy consumption per person in different countries*

MEDCs have had a bigger share of current world energy consumption. In the future, it is expected there will be more rapid growth in energy demand from countries that are now LEDCs. China and India lead the world's economic growth and increase in energy consumption.

Common error

Renewable and non-renewable energy sources are often confused.

Benefits of increased energy consumption	Disadvantages of increased energy consumption
Electricity makes the daily household tasks easier – it provides heat, light in the evenings, television and computers. People do not have to collect fuel or use candles or lamps at night.	We may be using non-renewable energy sources too quickly and face running out of supplies.
Modern transport systems are mainly based on oil (petroleum) and allow goods to be moved around the globe and people to travel.	The use of **fossil fuels** is resulting in air pollution and causing global warming.
Industry requires energy to make it work. Without it the economy cannot grow, wealth cannot be increased and people's lives will not be improved.	The interdependence of countries on each other for energy supplies, e.g. oil and gas, can lead to conflicts.

Exam tip

Make sure that you know a case study of energy supply in a country or area.

The benefits and disadvantages of energy sources

Energy source	Benefits	Disadvantages
Coal	Many countries still have large reserves of fuel, for example coal in South Africa and Germany.	Power stations are major sources of carbon dioxide, nitrogen oxides and sulfur dioxide which are health hazards and cause global warming. The air pollution caused by one country is often 'exported' to another by prevailing winds.
		World coal reserves may only last for another 300 years.
		Coal is bulky and normally needs to be transported by rail.
Oil (these also apply to natural gas)	It is easy to transport by pipeline or bulk tanker.	Burning oil produces greenhouse gases and leads to global warming.
	It is the only fuel in mass use for motor vehicles, although this may change.	Oil spills from leaking tankers and pipelines can cause pollution which kills wildlife.
	It is also a raw material in the chemical industry.	World oil production is concentrated in a few countries, which control the supply and prices.
Natural gas	Electricity generation is less expensive with natural gas than with oil. Gas-fired power stations are less expensive to build than plants that use coal, nuclear or most renewable energy sources.	
	Gas is 'cleaner' than oil.	
Nuclear power	Only very small amounts of uranium are needed to produce large amounts of energy.	There have been serious incidents at nuclear sites which have led to leaks of radioactivity. Radioactivity is a known cause of diseases, such as cancer and leukaemia.
	Uranium ore will not run out in the foreseeable future (some people classify nuclear power as renewable).	The cost of shutting down old nuclear plants (decommissioning) is very high.
	It does not produce greenhouse gases and acid rain.	The radioactive waste from power stations remains a health hazard for hundreds of thousands of years. It requires careful storage and is difficult to dispose of safely.
	The safety record of nuclear power stations has improved and the industry is highly regulated in many countries.	The capital costs of building nuclear power stations are extremely high.

Exam tip

Learn the benefits and disadvantages of each energy source.

Exam tip

Learn how burning fossil fuels increase the greenhouse effect and leads to global warming.

Energy source	Benefits	Disadvantages
Fuelwood in LEDCs	It may be 'free' to the user and so it provides an accessible source of fuel for heating and cooking for poor people.	In some areas, natural woodland is being cut quickly. This means that longer and longer distances have to be walked to collect wood, meaning a lot of hard work and more time being taken up.
	If there is enough land then wood can be a renewable, sustainable energy source.	Deforestation may lead to exhaustion of soils and soil erosion, so that the forest cannot grow back.
	For rural people who live near towns, surplus wood is often collected and sold to townspeople, sometimes by the roadside, so it becomes a cash crop.	Burning wood in confined spaces on inefficient stoves leads to respiratory illnesses.
	It does not require high-tech equipment.	
Geothermal power	It is extremely cheap and reduces dependence on fossil fuels.	It is restricted to areas with suitable geology.
	It does not produce greenhouse gases.	Areas with suitable geology are sometimes affected by earthquakes and volcanoes.
	The water is pumped back into the ground and reused.	Although it is usually classified as renewable, each well may only be used for about 25 years.
	Unlike other types of renewable energy, it can operate at any time of the day and year and is not affected by the weather.	The groundwater is saline and often poisonous.
Wind power	It does not cause air pollution, global warming or acid rain.	Wind power cannot be used during calm periods or storms.
	It has very little effect on the local ecosystem, except very occasionally killing birds.	Many people consider wind farms to be a form of visual pollution, especially in areas of natural beauty.
	In Europe, winds are strongest in winter when there is a peak demand for electricity.	The technology is relatively new and at present very large numbers of turbines are needed to generate fairly modest amounts of electricity.
	After the initial capital input, production is cheap as the fuel is free.	
Hydroelectric power	Once a dam is constructed, electricity can be produced at a constant rate.	Dams are extremely expensive to build and they must operate for many decades to make a profit.
	The power stations can respond quickly to changing demand.	The flooding of large areas of land means that the natural environment is destroyed, along with natural habitats and historical or archaeological features.

Energy source	Benefits	Disadvantages
	There are no fuel costs.	People living in villages and towns that are in the valley to be flooded must move. In some countries, people are forcibly removed so that construction can go ahead.
	The lake that forms behind the dam can be used for water sports and leisure activities.	Although modern planning and design of dams is good, in the past old dams have been known to collapse. This has led to deaths and flooding.
	The water can be used for irrigation and other purposes.	When a river flows from one country to another a dam in one country affects the flow of the same river in the next country and can lead to conflicts between neighbouring countries.
	There is no atmospheric pollution.	Dams catch sediment that would have flowed down the river and increased the fertility of soils when it settled from floodwaters. The sediment also reduces the capacity of the dams.
Tidal power	Tides are more predictable than the wind and the sun.	Relatively high cost of developing new technology.
	Tidal power is practically inexhaustible (a renewable resource).	Limited availability of sites with sufficiently high tidal ranges or flow velocities.
		Some types involve the creation of lakes or dams which disrupt existing ecosystems.
Solar power	It is safe and pollution-free.	The initial capital input is high.
	After the initial capital input, production is cheap as the fuel is free.	It is not as effective in cloudy countries.
	It can be used effectively for low-power uses, such as heating swimming pools or central heating.	It is less effective in high-latitude countries, where more power is needed in winter but the days are shorter and the sun is lower in the sky, giving less light.
	Its greatest potential is in warm, sunny countries or in LEDCs, where people live in locations which are isolated from the national electricity grids.	It is less effective for high-output uses, such as powering colour TVs.
Biofuels	Prices could be more stable than world oil prices.	Land previously used for the production of food has been changed to produce crops for biofuel production, leading to increases in world food prices and decreases in the food supply.
	Supplies can be more secure and reduce reliance on imported fuels.	
	Fewer pollutants are produced than by than fossil fuels.	
	It is carbon-neutral, in that the growing crop absorbs carbon dioxide from the air which balances emissions from the burnt fuel.	

Wave power

Energy can be captured from waves in the sea. Unlike tidal power, wave energy is not constant and depends particularly on wind strength. Wave power is not used commercially today.

Exam tip

Questions on energy in Paper 2 often involve graph reading.

Recap

- World energy consumption is increasing, especially in LEDCs.
- Generally, MEDCs use much more energy than LEDCs.
- Coal, oil and natural gas are non-renewable energy sources.
- Geothermal, wind, hydroelectric power, wave and tidal power, solar power and biofuels* are renewable energy sources.
- Each different energy supply has its own benefits and disadvantages.

*biofuels including fuelwood are only renewable if replacement planting goes on.

Apply

(a) Make a list to describe the expected changes in energy consumption shown in these graphs.

(b) Make a list of possible reasons for these changes.

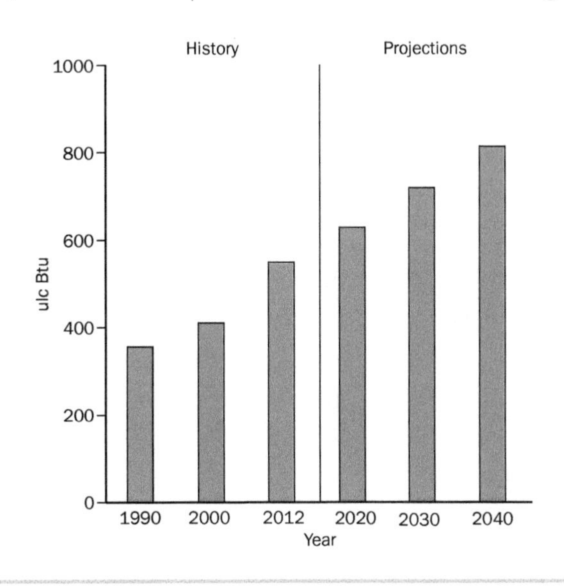

See www.oxfordsecondary.com/esg-for-caie-igcse for the answers to the 'Apply' task.

Review

Non-renewable	Renewable	Nuclear
coal oil natural gas	geothermal wind hydroelectric power wave and tidal power solar power biofuels*	The IGCSE syllabus does not classify nuclear power as either renewable or non-renewable

*biofuels including fuelwood are only renewable if replacement planting goes on.

Sample question

1. For a country or area that you have studied, describe and give reasons for its energy supply. **[7 marks]**

Analysis

✓ It would be better to choose a country or area where energy is from more than one source.

✓ Remember to give reasons for the choice of energy source.

Student answer

Germany uses a variety of different energy sources. It has traditionally generated electricity from coal. This includes lignite, e.g. mined in Sachsen, and bituminous coal mined in Nordrhein-Westfalen (the Ruhr Coalfield). Power stations are located close to the coalfields to reduce the transport cost of coal. Some power stations are on the coast and on navigable rivers (e.g. the Rostock Power Station) and use imported coal. Germany recognizes that its thermal power stations give out large volumes of carbon dioxide, a greenhouse gas.

There are 17 nuclear power stations distributed throughout the country. Many are located near large rivers to supply cooling water. In 2011, the government announced that it would close all nuclear power stations in the country by 2022. Public opinion is often divided on the issue. If Germany were to phase out nuclear power production and continue to reduce its carbon emissions, it would need to import large amounts of electricity. This would make Germany dependent on neighbours, such as France, for electricity (much of which would be produced by nuclear power stations).

Germany intends to cut overall electricity consumption by 10% and the use of coal and renewables would be increased. Germany may also want to reduce its dependence on gas imported from Russia.

[Level 3: 7 out of 7]

See www.oxfordsecondary.com/esg-for-caie-igcse for the mark scheme for this question.

Examiner feedback

This is an excellent answer. It discusses different sources of energy and the changes in government policy and the reasons for changes. Locations of power stations are discussed and specific locational information is given. The style of an answer to this question will partly depend on the example chosen.

Key ideas

- **Water is extracted from rivers, wells and boreholes.**
- **Water consumption varies from country-to-country, depending on the level of economic development.**
- **Water use for agriculture (irrigation) is greater than domestic or industrial use.**
- **A variety of factors can lead to water shortages.**
- **Water shortages affect people, the environment and the economy.**

Methods of water supply

Water supply comes from two main types of source, both of which come from rainfall (or snow melt). These are **surface water** and **groundwater**.

It is possible to **desalinate** water. However, desalinating seawater uses a lot of energy. It is regarded as a solution to the lack of freshwater in rich countries with dry climates, such as Australia, Saudi Arabia and the United Arab Emirates. Kuwait produces all of its drinking water by desalination.

Exam tip

Make sure that you know a case study of water supply in a country or area.

Uses of water

Water is in demand for:

- agriculture – irrigation in dry areas
- domestic use – in MEDCs people use large volumes of water each day for washing, flushing toilets, watering gardens and even washing cars. In many LEDCs, this luxury is not available
- industrial use – many industries use large volumes of water in processing (e.g. paper manufacture) and cooling (e.g. power stations).

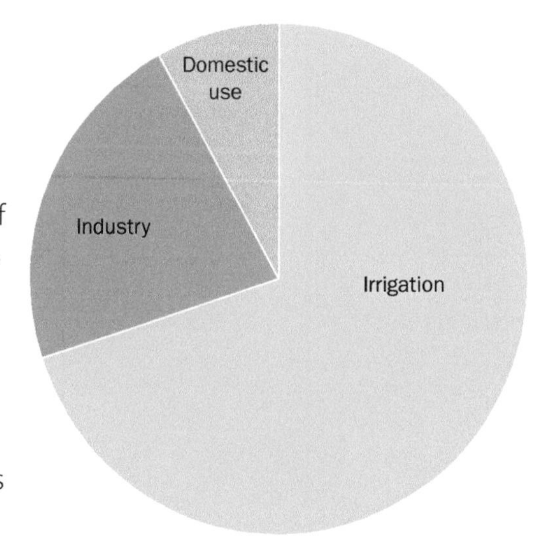

Fig. 25.1 *Breakdown of world freshwater consumption*

Common error

People often think that industry is the world's biggest consumer of water. Agriculture is easily the biggest.

Exam tip

Examination questions on Paper 1 are often about water supply and hydroelectric power combined. Your case study might cover both topics.

Common error

Water deficits don't just occur in LEDCs.

Water shortages

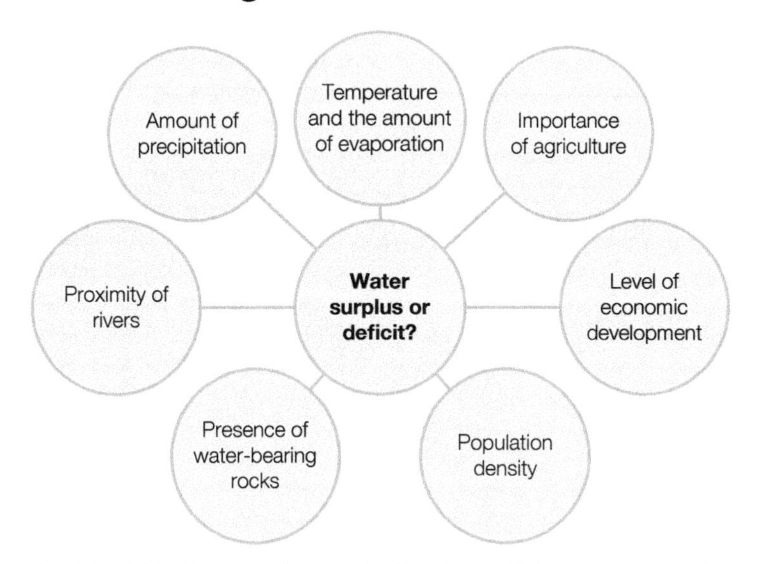

Fig. 25.2 *Factors affecting whether there will be a* ***water surplus*** *or* ***deficit***

Impact of water shortages

Water supplies should be sustainable – they should not run out and their extraction should not endanger biodiversity.

- Social impact: It is important that water supplies should be clean. Where people consume infected water they are prone to diseases, such as cholera and diarrhoea. This is a major cause of death in children in some LEDCs.
- Environmental impact: Major dam/**reservoir** construction projects raise economic, social and environmental issues, by lowering river levels so that local plants and animals cannot survive.
- Economic impact: When rivers flow across international boundaries, the use of water by one country can affect other countries downstream. Lack of clean water hinders economic development. Where there is insufficient irrigation water, agriculture industry suffers.

Recap

- Water supply includes surface water (rivers and reservoirs), groundwater (wells and boreholes) and, in rare cases, desalination of sea water.
- MEDCs use a lot more water than LEDCs, although there are big variations from country to country.
- Water use for agriculture (irrigation) is greater than domestic or industrial use.
- Water shortages are caused by a variety of physical and human factors.
- Water shortages affect people, the environment and the economy.

Apply

For an area that you have studied, make a list of the advantages of the area for supplying water and a list of the impacts of water supply schemes on the area.

See www.oxfordsecondary.com/esg-for-caie-igcse for the answer to the 'Apply' task.

Review

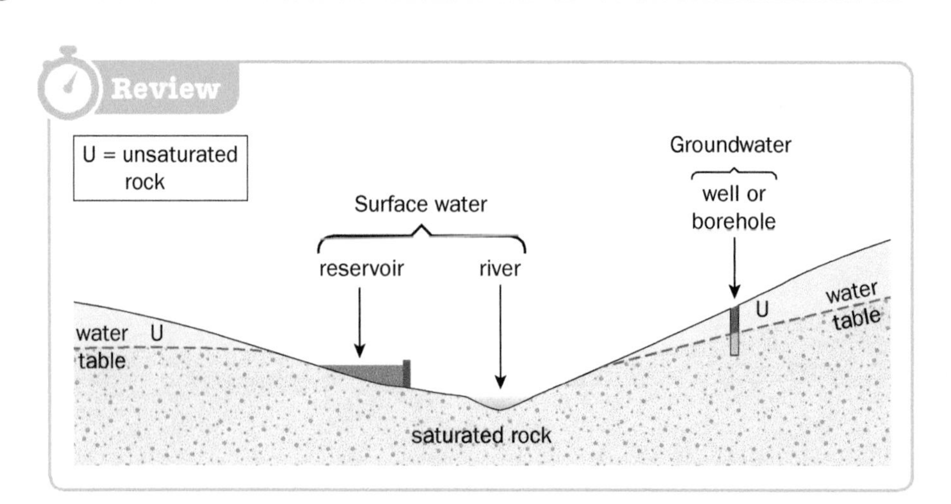

Raise your grade

Sample question

1. Study the map showing regions of water surplus and water deficit.

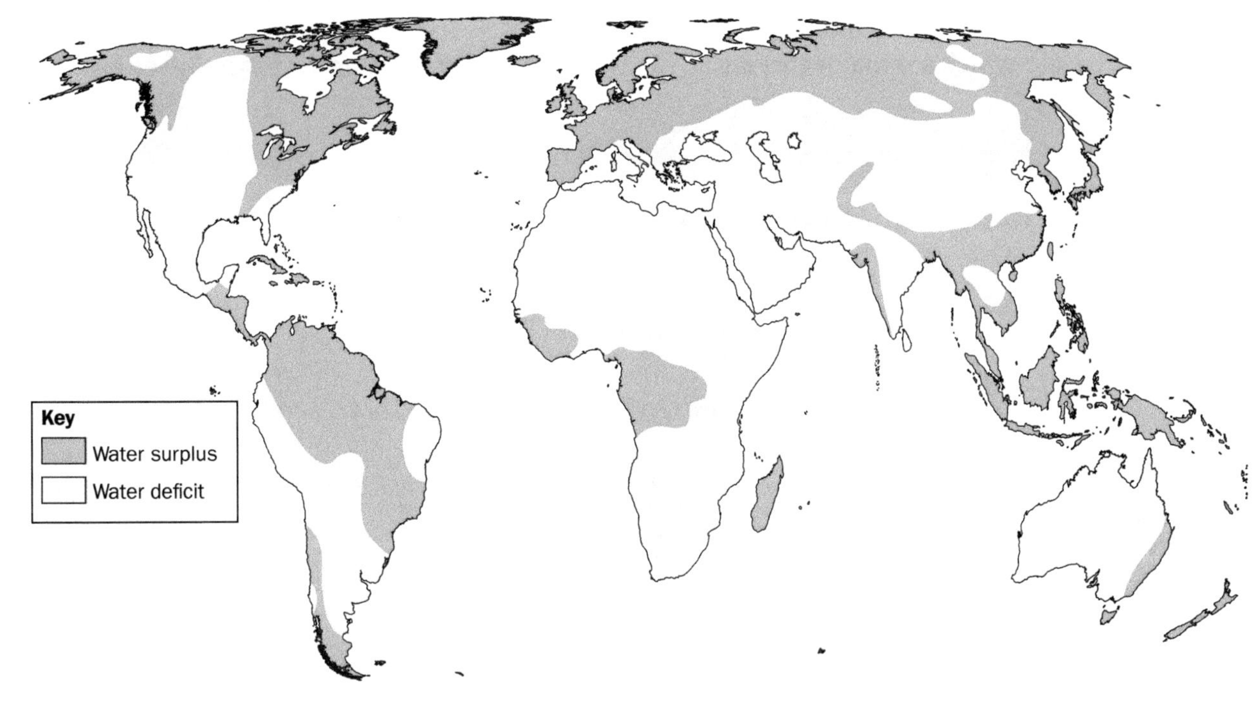

(a) What is meant by the term *water deficit*? **[1 mark]**

(b) Give **two** causes of water deficits. **[2 marks]**

(c) Are the areas with water deficits all LEDCs? Support your answer with evidence from the map. **[3 marks]**

Analysis

✓ All parts of the question require you to use your own knowledge, but you must also use the map in part (c).

✓ Part (c) has three marks. Make sure you give enough evidence to score all these marks.

Student answer

(a) Dry areas with not enough rainfall. [0 marks]

(b) Very low rainfall and hot climates. [1 mark]

(c) Not all areas with water deficits are LEDCs. For example, countries like Italy are MEDCs but have water deficits. [2 marks]

Total: 3 out of 6

Examiner feedback

In part (a), the definition is not correct; it must refer to the demand for water. In part (b), a mark has been scored for low rainfall, but it has not been explained that high temperatures lead to loss through evaporation, so no mark can be awarded for this point. In part (c), there is a correct conclusion, but only one piece of evidence (Italy). If there was another example, such as Australia, then full marks would have been scored.

See www.oxfordsecondary.com/esg-for-caie-igcse for the mark scheme for this question.

Key ideas

- Economic activities threaten the natural environment and people both locally and globally.
- Poor farming techniques lead to soil erosion and desertification, both of which are major threats.
- Soil is a very important resource, as it allows crops and vegetation to grow, which are necessary for human and animal survival.
- Soil takes a very long time to form, but can be very rapidly removed by erosion.

Soil erosion

Soil erosion occurs when:

- the soil, damaged by poor agricultural practices, loses its **structure** and becomes loose
- the soil is exposed to wind and rain and not covered by vegetation or crops.

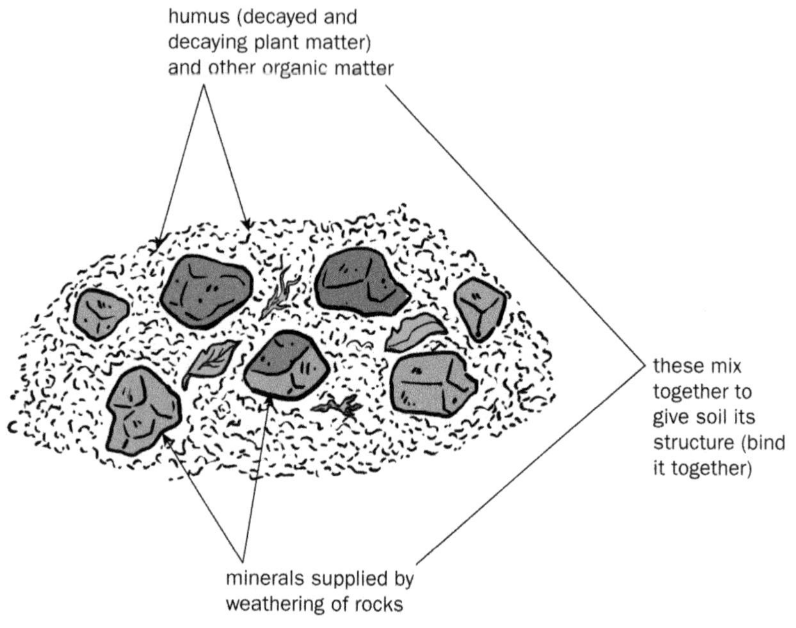

Fig. 26.1 *Soil contains tiny rock fragments (minerals) held together by decayed organic matter. Both these supply plant foods (nutrients).*

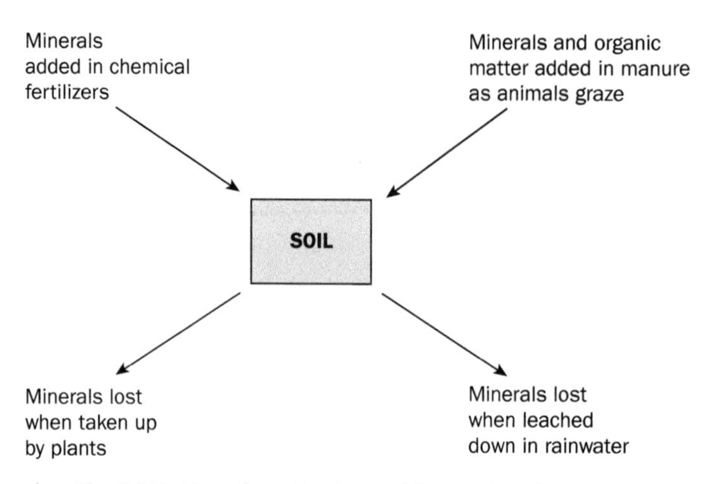

Fig. 26.2 *How the soil gains and loses minerals*

Soil erosion by wind

Loose soils with poor structure are easily removed by strong winds, especially when the soil dries out after periods of little rain. Dry soil particles are lighter and more easily picked up by wind.

Soil erosion by running water

This occurs on slopes when rainfall is so heavy that all of it does not soak into the ground and surface runoff occurs down the slopes. Soil is removed either in **sheet wash** on gentler slopes or by **gullying** after rainstorms on steeper slopes. It is often started by poor farming methods and occurs most rapidly at the end of the dry season when the soil is exposed.

Poor agricultural practices leading to soil erosion

- **Monoculture** damages the soil by removing the same **nutrients** all the time.
- **Overcultivation (overcropping)** depletes the soil's nutrients.

Both the above farming practices cause the soil to lose structure and loosen by exhausting its nutrients.

- **Overstocking** – keeping more animals than the land can support – leads to the death of plants by trampling, and **overgrazing** leaves the land bare.
- Cultivation on steep slopes and ploughing up and down the slope encourage rain to wash soil down the slope as the furrows channel the run off.
- Cultivation in dry areas allows erosion when soils are bare after harvesting.
- Leaving land fallow (unused) after harvesting leaves it unprotected from wind and rain erosion.
- Removal of vegetation to farm in unsuitable areas exposes the soil to erosion.

Vegetation cover:

- provides humus to maintain soil structure by binding the particles together
- gives a protective cover to protect the soil from erosion
- intercepts heavy rainfall and reduces rainsplash erosion
- holds the soil in place with its roots.

Common error

Remember that dry soils are not infertile. Dry soils need water for plants to grow. Fertility means the soil has insufficient *nutrients* for crop growth.

Common error

Soil erosion is loss of soil in wind or rainwater, not its removal by streams.

Desertification

Poor land use practices and drought set in motion a chain of physical processes which lead to desertification, especially in semi-arid lands with fragile ecosystems and population pressure.

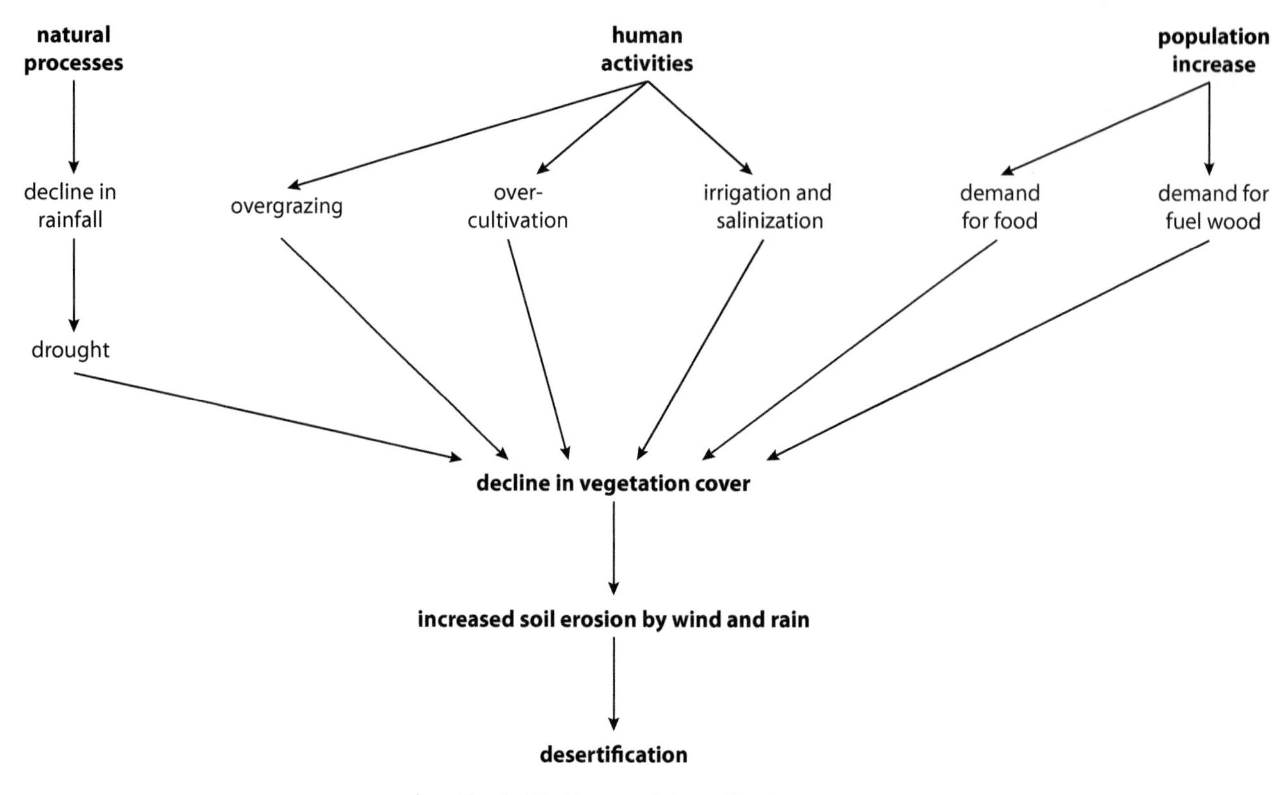

Fig. 26.3 *Causes of desertification*

Common error

Desertification is not the spread of a desert. Many desertified areas are far away from hot deserts. Land can recover from frequent droughts if it is well managed.

Population growth causes the excessive gathering of fuelwood, which destroys protective vegetation. It also increases the need to produce more food by:

- cultivating the land more intensively
- keeping more animals
- clearing more vegetation to cultivate more marginal land.

Deforestation may lead to both exhaustion and erosion of soils so that forest cannot regrow.

Exam tip

Be able to locate an area where soil erosion and/or desertification is occurring and to explain its causes and describe its effects. Also, be familiar with ways in which the problem could have been avoided.

Key ideas

- Development should be managed so that it is sustainable.
- It is important to conserve resources, many of which are finite.
- Soil can be considered to be finite because it takes so long to form.
- Conservation of soil is vital to grow food for the increasing population.
- Attempts to increase agricultural production and food supply must be sustainable.

Sustainable management

Soil conservation methods

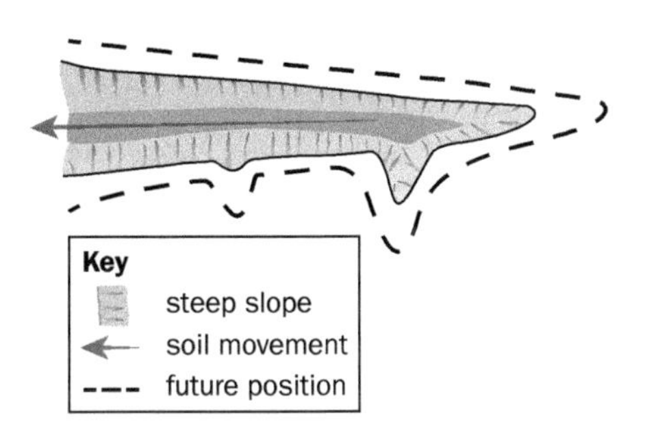

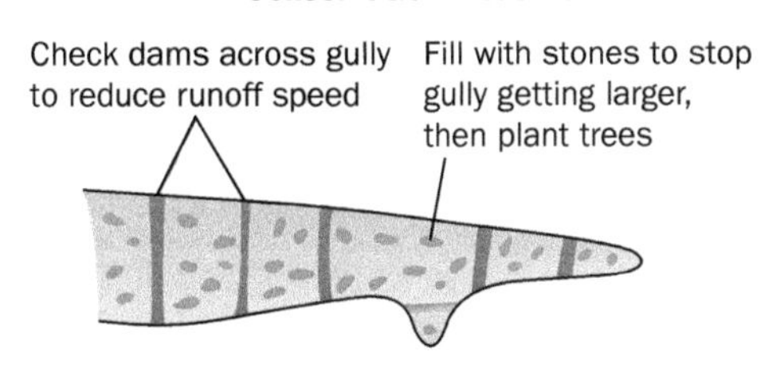

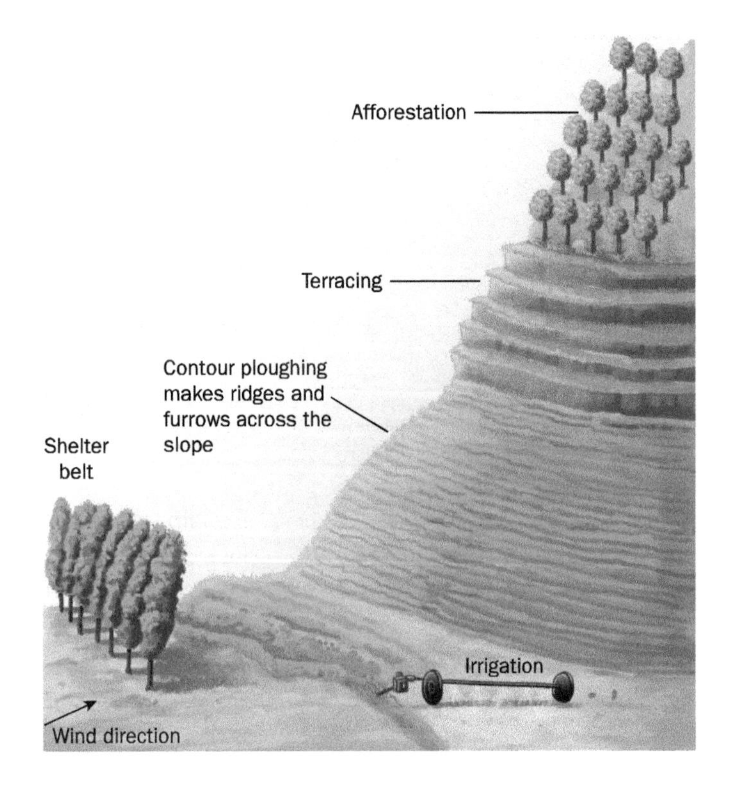

Fig. 26.4 *Some soil conservation methods*

Other strategies for sustainable agriculture

- Use organic fertilizer, such as animal manure. Artificial fertilizer does not provide organic matter to maintain the soil structure. Animals can be put to graze on land where crops will be grown so they fertilize the soil. Manure collected from animal sheds can be spread on the fields. Crop residues left after harvesting can be ploughed back into the soil. Cover cropping also provides organic matter. Educate people who use manure as fuel to use it on their soil instead.
- Limit livestock numbers to the carrying capacity of the land and reduce stocks in drier years. This prevents overgrazing.
- Rotate livestock regularly between different pastures. The land needs to be fenced to enable this.

Crop rotation: Different cops are grown in the same field for 3–4 years before the first crop is grown again. As they take different nutrients from the soil, it does not become exhausted, lose structure and become loose and easily eroded.
Strip cultivation and inter-cropping: Narrow bands of crops grown at right angles to the prevailing wind. Different crops are grown in between and harvested at different times, so the soil is never completely bare. This traps any wind-blown soil. **Cover cropping** (the planting of a fast-growing crop after the main crop has been harvested) is also used.
Dry farming methods (e.g. planting without ploughing, covering the soil with a mulch, planting crops in alternate years) aim to reduce water loss from the soil.

Fig. 26.5 *How some other soil conservation methods work*

- Soil is a vital resource that must be conserved.
- It takes a very long time to form, yet is being degraded and lost very quickly by human mismanagement, particularly by damaging deforestation methods and poor farming practices.
- Because of the above, there is less natural vegetation cover and habitat for wild animals, less pasture for farm animals and a reduction in crop production. Food supplies are being reduced at a time when the population is increasing quickly in many parts of the world.
- Desertification is the final result of human mismanagement and many areas are at severe risk. Natural processes, such as droughts and climate change, may also cause desertification.
- Soil must be conserved by sustainable management methods to conserve the natural environment and the health and lives of people.

See **www.oxfordsecondary.com/esg-for-caie-igcse** for the answers to the 'Apply' task.

Apply

Explain why each of the following could contribute to soil erosion: lack of education; poverty; lack of choice; population growth; lack of government control; the desire of big companies for profit.

Review

The main focus in this chapter has been the environmental risks of poor farming practices and how farming can be made more sustainable. But it is very important to be aware of the damaging destruction of forests that is occurring. Many forests are shrinking at an alarming rate with extremely serious consequences that are likely to affect the whole world.

Sustainable methods of land use that could replace damaging methods of deforestation include:

- selective logging of forests and replanting instead of clear cutting
- afforestation, which reduces all soil erosion types to a minimum.

There is a case study of deforestation and attempts to manage the rainforests in Borneo in Chapter 7 of *Complete Geography for Cambridge IGCSE and O Level*. In Chapter 9 of *Complete Geography for Cambridge IGCSE and O Level*, the case studies of agriculture in Brazil, in the Prairies of Canada, and in Swaziland all include information about soil erosion.

Sample question

1. Study the diagram of agricultural land in a mountainous area.

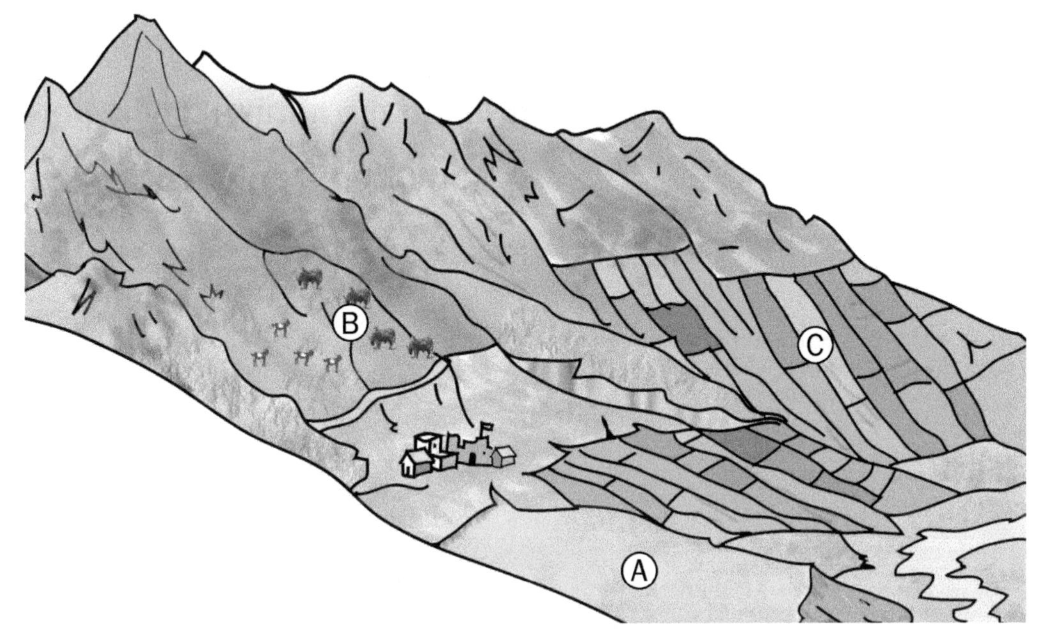

(a) Explain why area A could be badly affected by wind erosion. **[1 mark]**

(b) Give **two** ways in which the farming type at B can cause soil erosion. **[2 marks]**

(c) Explain what will happen at C if it rains torrentially after the crops have been harvested. **[3 marks]**

Analysis

✓ The question illustrates the problems that occur in mountainous areas if poor farming practices are used.

Student answer

(a) The land appears to have been left fallow which means the wind can remove the bare soil. [1 mark]

(b) If more animals are kept than the carrying capacity of the land, overgrazing will occur. [1 mark]

(c) The rain will wash the bare soil down the slope into the river. This will be made worse if the farmer ploughed the field up and down the slope, because the furrow will act as a channel for the rain to run down. [3 marks]

Total: 5 out of 6

Examiner feedback

The answer to part (a) is correct. Part (b) is awarded one mark for the first point in the mark scheme, but a second way is required for the second mark. Part (c) is a well explained answer with three clearly made points.

See www.oxfordsecondary.com/esg-for-caie-igcse for the mark scheme for this question.

Key ideas

- **Pollution of all types has already had an adverse impact on the natural environment and people.**
- **Developments in industry, food production, energy, tourism and transport have all contributed to an increase in pollution.**

Common error

Remember that the transfer of heat from ground to cloud and back to ground that is involved in both the natural and enhanced greenhouse effect is done by radiation. (Reflected solar radiation passes through the atmospheric gases.)

The impact of pollution on the natural environment

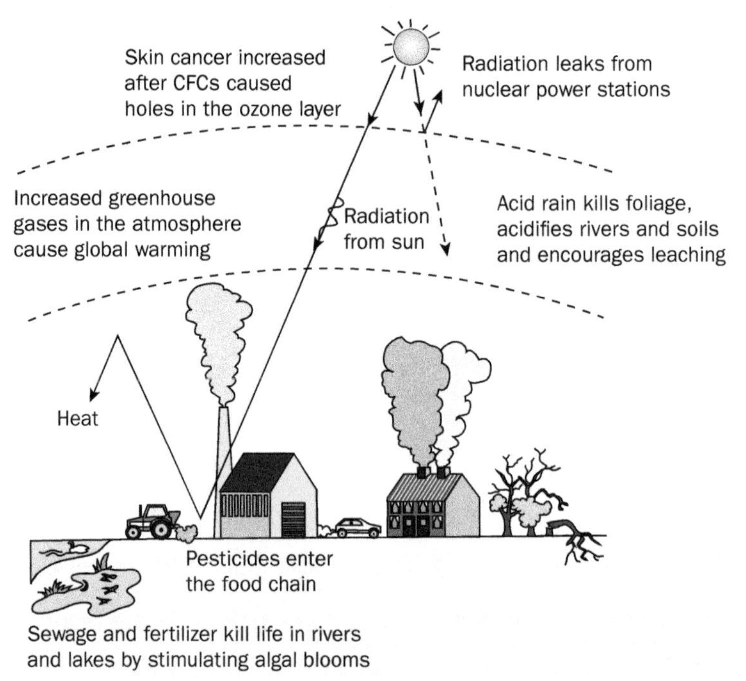

Fig. 27.1 *Sources and impacts of pollution*

Air pollution

It is in the major cities of LEDCs and NICs that the highest levels of air pollution occur, especially in countries where industrialization is rapid, such as China and India. However, cities in MEDCs are also affected by a brown fog, **photochemical smog**. It contains air pollutants from industries such as oil refining, gas production, chemical manufacturing and power stations and, in particular, gaseous hydrocarbons and nitric oxides from vehicle exhausts. Vehicles are the main source of pollution in cities.

These pollutants put human health at risk by causing breathing problems from reduced lung function and are believed to worsen heart problems. They also irritate the eyes, nose and throat. Particulates from diesel exhausts are known to be associated with COPD (Chronic Obstructive Pulmonary Disease) and lung cancer.

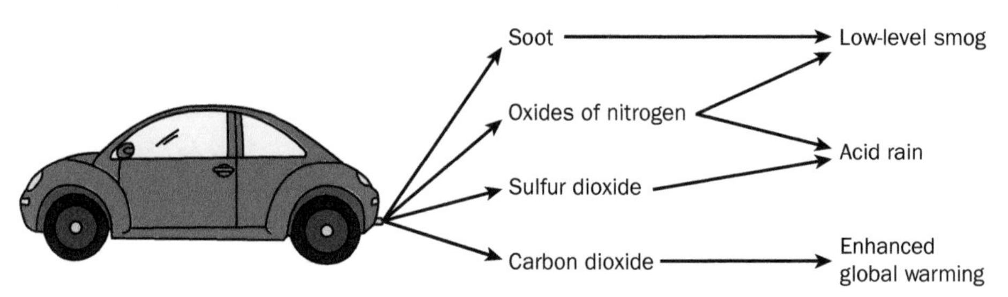

Fig. 27.2 *Air pollution from vehicle exhausts has major consequences for human, plant and animal life*

Water pollution

Source of pollutants	Examples
Agriculture	Animal slurry, fertilizers, pesticides
Domestic waste	Organic waste
Industrial processes	Oxides of nitrogen, sulfur dioxide, toxic heavy metals
Sewage	Organic matter, phosphates from detergents

In addition to acid rain and eutrophication (algal blooms caused by fertilizers), aquatic life and water supplies are threatened in other ways.

- In areas where there is no sewage system, such as some shanty settlements in LEDCs, raw sewage is simply dumped or left in open drains. In other areas, raw sewage is not treated before emptying from sewers into rivers or the sea.
- Some liquid industrial waste is discharged into rivers from factories. Toxic heavy metals enter the food chain through fish and eventually kill the fish. Fifty years ago many of the urban rivers of Europe had no life and many in LEDCs are now like that.
- In some LEDCs, solid domestic waste, often including battery acid, is dumped in rivers.
- Even high-tech industries pollute the environment. Accidental spills and leaks of solvents and acids can cause toxic substances to pollute both the air and water.
- There is now a lot of concern about microplastic in our waters, which is so small it is eaten by aquatic animals and moves up the food chain to humans. Plastic items are now discarded in vast amounts, litter often floats down rivers to the sea. It degrades very slowly.
- Oil spills from tankers and pipelines cause river and sea pollution which kills wildlife. The marine ecosystem is being damaged by shipping, especially ships carrying oil and oil products in and out of ports. A lot of oily discharges are pumped out within port areas in LEDCs.

In LEDCs, water courses and wells used for drinking water are often contaminated with sewage and waste water, leading to diseases such as cholera and typhoid.

Visual pollution

Many examples of visual pollution exist. Power stations, factories and large concrete buildings are ugly. Quarries scar the landscape in rural areas. Deep mines have unsightly mineral storage and waste heaps, while opencast mining has enormous open pits and heaps of stored soil and rock.

Noise pollution

Sources of noise pollution include:

- blasting and the use of large-scale machinery in quarries and opencast mines
- vehicles, trains and aircraft taking off and landing
- factories
- congregations of large numbers of people, e.g. at sports fixtures.

The problems are worst where cities have grown rapidly without proper planning. Noise pollution can be bad for human health if it affects sleep patterns and can be seriously disturbing for wildlife.

Common error

Remember not to refer just to 'pollution'; it is too vague. Always specify the *type* of pollution you are writing about.

Key ideas

- It is vital for the future health of the world and its people that pollution of all types is controlled.
- In MEDCs, strict laws ensure that dangerous waste does not normally enter the air, seas or rivers or contaminate the land.
- In many LEDCs, pollution prevention measures are often too costly to adopt.
- In the natural greenhouse effect of the Earth's atmosphere, carbon dioxide produced by humans and animals was balanced by the amount of carbon dioxide taken in by vegetation, especially trees.
- That balance no longer exists because of the greater output of greenhouse gases by increased industrialization, transport and forest clearance.
- Carbon dioxide is the greenhouse gas of greatest concern.
- Warming causes greater evaporation of water, adding more water vapour, a greenhouse gas, to the atmosphere, which causes even more warming.
- There has been a 0.9 °C increase in the average world temperature since 1950.

Managing economic development to make it more sustainable

Air pollution

- Government agencies set standards for air quality which are enforced by law.
- National and international agencies are driving the reduction in the production of carbon dioxide (the **carbon footprint**) by, for example, using alternatives to fossil fuels, using energy more efficiently and reducing energy consumption. People are encouraged to use public transport, walk or cycle. Taxes are imposed on the use of energy made from fossil fuels and on emissions produced by them.
- In MEDCs, strict regulation of vehicles and industrial plants has greatly reduced air pollution. Reducing traffic congestion also helps. Catalytic converters are used in petrol and diesel engines and in stoves and heaters that use kerosene to reduce the toxicity of their emissions.
- Many modern industries use **scrubbers** to remove harmful gases, such as sulfur dioxide and carbon dioxide.
- In LEDCs, new fuel-efficient stoves which cause less smoke are helping reduce pollution from cooking using fuelwood.

Additional information about the enhanced greenhouse effect

In response to initiatives by the IPCC (Intergovernmental Panel on Climate Change) many countries signed agreements to reduce their greenhouse gas emissions but many have not met their targets. Naturally, NICs were unwilling to reduce their increased industrialization based on fossil fuels as the developed nations had already built their stronger economies in that way. The good news is that many NICs and LEDCs are now installing alternative energy, as more advanced technology is making it cheaper, while still using fossil fuels.

Meanwhile, the carbon dioxide content of the atmosphere continues to rise and peoples like the Inuit of the Arctic, see their environment changing. It is becoming impossible for them to continue successfully with their traditional way of life, hunting polar bears and seals on the sea ice and catching fish through holes in the ice.

Water pollution

In MEDCs and many towns and cities in LEDCs:

- strict regulation of industry has improved river water quality and fish have returned to many rivers
- sewage systems treat sewage to make it safe before it is released into rivers or seas
- water treatment plants produce potable water.

Treatment plants are still not available everywhere in LEDCs, especially in rural areas where villagers often have to rely on wells.

It is vital that substitute materials are used for packaging and bags instead of plastic, wherever possible.

Exam tip

Be able to name an example area where economic development has taken place and caused the environment to be at risk from pollution.

Methods of cleaning up oil spills in the ocean		
Boom	**Detergent spray**	**Skimmer**
Floating inflatable tubes prevent slicks from spreading	Chemicals break up oil into droplets, dispersing larger slicks	Oil drawn up absorbent belt; rollers scrape and squeeze oil into tank

Visual pollution

In MEDCs, ugly developments are often screened from view by trees or fencing and town planning keeps unsightly industrial developments out of sight of areas of better class housing. The walls of some tall buildings are camouflaged with painted patterns.

Noise pollution

Some countries have laws which limit noise from factories and homes. Also, noisy industrial sites are placed away from homes.

27 Environmental risks of economic development (2)

Recap

Industrial development threatens lives and the sustainability of the natural environment. MEDCs have strict laws to reduce pollution, but laws are less stringent elsewhere. Industries depend on energy and transport, which can be high polluters.

- Nuclear power produces highly radioactive waste which remains dangerous for thousands of years.
- Fossil fuels produce carbon dioxide (a greenhouse gas) and sulfur dioxide, which acidifies rain. Diesel fuel use causes lung diseases.
- Toxic heavy metals from industry get into the food chain in rivers and seas. They build up in the bodies of consumers and are at their most concentrated at the top of the food chain; people are often at the top.
- Increasing agricultural output relies on using pesticides, fertilizers and other polluting chemicals.

Apply

With reference to a named industry explain how its development has caused serious harm to the environment and people of the area. Prepare a list of points to include in answer to this question.

See www.oxfordsecondary.com/esg-for-caie-igcse for the answer to this 'Apply' task.

Review

- Pollution's increasing threat is a world problem as it sometimes spreads beyond the polluting country.
- Waste entering rivers can reach the interconnecting oceans and circulate to affect wildlife and fishing grounds far away from the pollution source.
- Atmospheric pollutants cross borders. For instance, radioactive substances from the 1986 nuclear power plant explosion at Chernobyl (in the Ukraine) were blown westwards to affect countries as far away as the UK. Polluted rain contaminated pastures and soils, and farm products were too contaminated to sell.
- International agreements to reduce pollution are vital. Everyone needs to be educated about the seriousness of the threat and how they can reduce their pollution footprints.

There are case studies of pollution on pages 67–73, 77, 263–66 and 294–95 of *Complete Geography for Cambridge IGCSE and O Level*. Also, the website has information in Question 1 for Chapter 10 about the Pearl River Delta industrial region

Sample question

1. Study diagrams A and B, showing the greenhouse effect and the enhanced greenhouse effect.

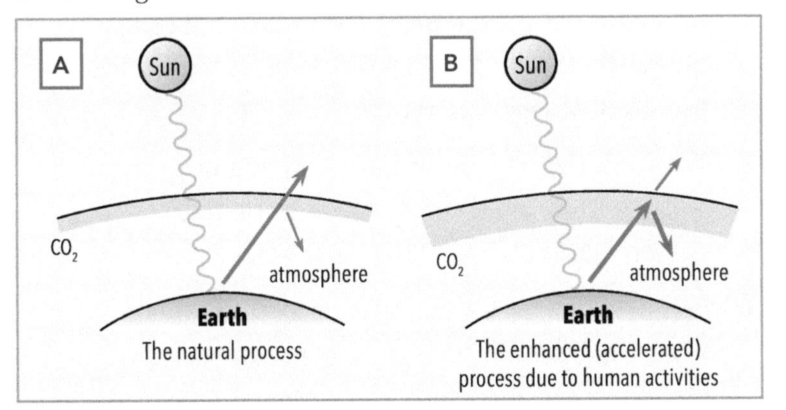

(a) What is the natural greenhouse effect, shown in diagram A? **[1 mark]**

(b) Explain how the enhanced greenhouse effect makes the atmosphere warmer. **[2 marks]**

(c) Describe the likely effects of global warming on the lives of people who live at the coast. **[3 marks]**

Analysis

✓ The Earth's atmosphere acts like a greenhouse – it lets solar radiation through but prevents much of the heat the Earth gains and re-radiates from passing through it back to space.

✓ Carbon dioxide occurs naturally in the atmosphere and is one of the greenhouse gases that kept our air and planet warm enough for life, before pollutants entered the air because of human activity. Pollution adds to it.

✓ Be very careful to use the correct terminology when explaining the natural and enhanced greenhouse effect. Greenhouse gases absorb the Earth's *radiation*, not the sun's rays that have *reflected* off the Earth's surface.

✓ Remember that global warming, or climate change as it is now more commonly known, changes the sea temperature as well as the air temperature and this may cause other changes, such as to pressure systems and ocean currents.

✓ The command word 'describe' means that you should state what the effects on the lives will be. Part (c) does not ask you to explain the changes.

Student answer

See www.oxfordsecondary.com/esg-for-caie-igcse for the mark scheme for this question.

(a) The greenhouse effect keeps the atmosphere near the Earth warm by letting solar radiation through to heat the Earth, but stopping some of the Earth's heat from escaping to space. [1 mark]

(b) The sun's rays reflect off the Earth and many are stopped from going straight back to space by carbon dioxide. [0 marks]

(c) Glaciers and sea ice will melt, causing sea levels to rise and flood their homes. Some low islands may disappear altogether, so people will have to move. The warmer sea will expand, adding to the sea level rise. [2 marks]

Total: 3 out of 6

Examiner feedback

The answer to part (a) includes the necessary points of what the greenhouse effect is, but part (b) incorrectly mentions reflected rays and also fails to state why more heat is kept in the atmosphere, so no marks are awarded.

The answer to part (c) would waste time because it also explains, which the question does not ask for.

Geographical skills

Chapters 28 to 30 are designed to help you understand and be able to use the geographical skills needed to interpret maps, including survey maps, photographs, data tables and different types of graphs and diagrams.

Geographical skills are found in all the question papers, but they are particularly important in Paper 2, where very few marks are for knowledge. Paper 2 also includes the compulsory survey map question worth 20 marks, so it is important that you have good geographical skills.

28 Survey (topographic) maps

Key ideas

- Survey maps are large-scale maps.
- They show the surface features of an area including relief, drainage, land use, settlement and roads.
- Map reading is a compulsory element of Paper 2.

Exam tip

The compulsory survey maps question has one third of the marks for Paper 2. A poor answer to this question may mean a poor overall mark.

Exam tip

In recent years, survey map questions have used maps from a variety of different countries. All these countries use slightly different symbols, so it is important to always check the meaning of symbols using the map key.

Using the key and symbols

The positions of different features on a map are shown by symbols. Different countries use different symbols on their maps, so it is always best to check the meaning of a symbol using the **key**, which is a list of the meaning of each symbol, usually at the side or at the bottom of the map.

Map scale

The scale of a map shows how distance on the ground has been represented on the map. A **large-scale map** might show a small area such as a school or a village, whereas a **small-scale map** might show a whole country. Scale is shown by the **representative fraction** and the scale line.

Representative fraction

Representative fraction	Distance on map	Distance on ground
1:25 000	1 cm	25 000 cm
	4 cm	1 km
1:50 000	1 cm	50 000 cm
	2 cm	1 km

Scale line

SCALE 1:25000

▲ **Fig. 28.1** *Use the scale line at the bottom of the map extracts when measuring distances*

Distance measurement

Remember to use the edge of a sheet of paper and the scale line when measuring distances. Avoid calculations!

Common error

Exam questions often ask for a distance on the map to be measured. Candidates who try to calculate this using the representative fraction often get the calculation wrong. Use the scale line!

Eastings and northings

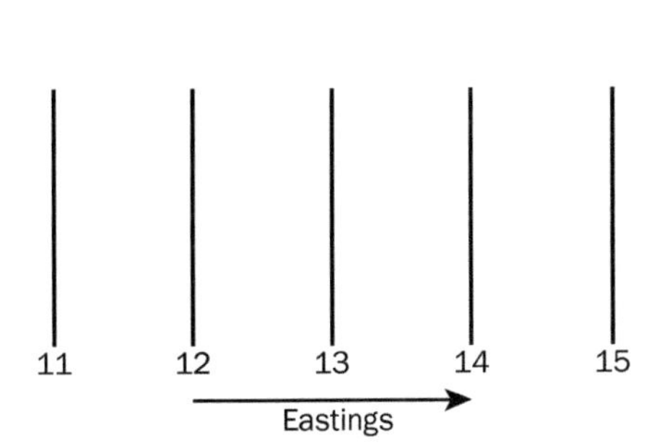

▲ **Fig. 28.2a** *Eastings run north-south and show far east a place is*

▲ **Fig. 28.2b** *Northings run east-west and show how far north a place is*

Four-figure grid references

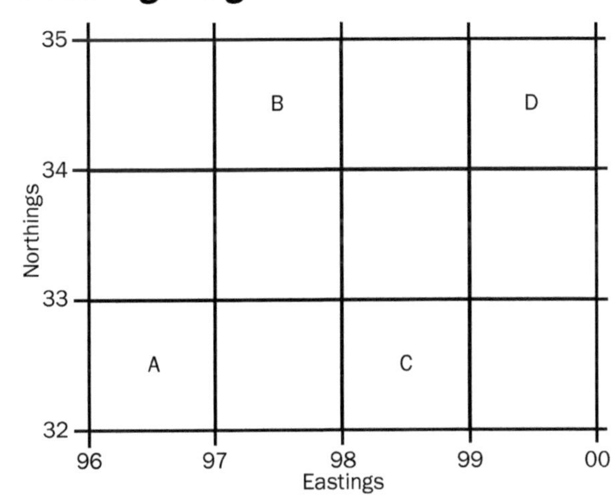

▲ **Fig. 28.3** *Four-figure grid references show the location of a whole grid square*

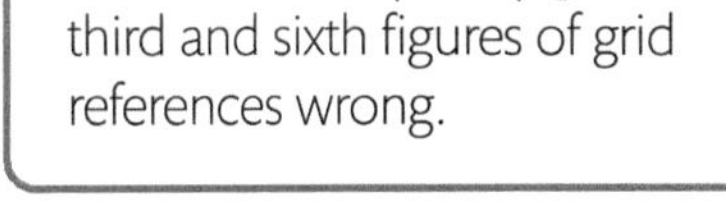

Common error

Candidates frequently get the third and sixth figures of grid references wrong.

The four-figure **grid reference** of A is 9632, B is 9734, C is 9832 and D is 9934.

Six-figure grid references

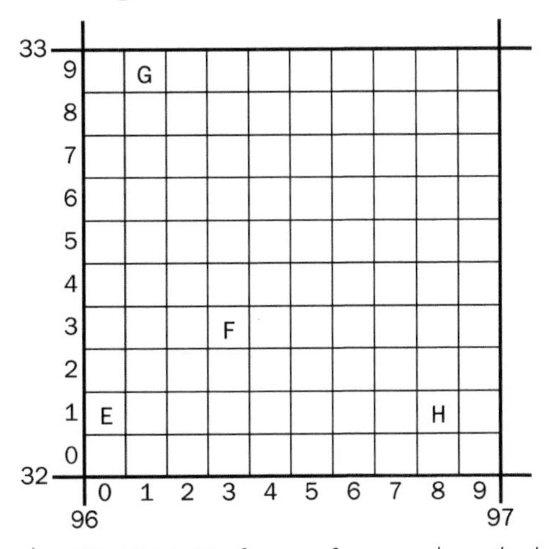

▲ **Fig. 28.4** *Six-figure references show the location of smaller features*

The six-figure grid reference of E is 960321, F is 963323, G is 961329 and H is 968321.

Exam tip

Use the method for measuring the third and sixth figures described opposite. Remember that anything in the first tenth is 0, and anything in the last tenth is 9.

Compass directions

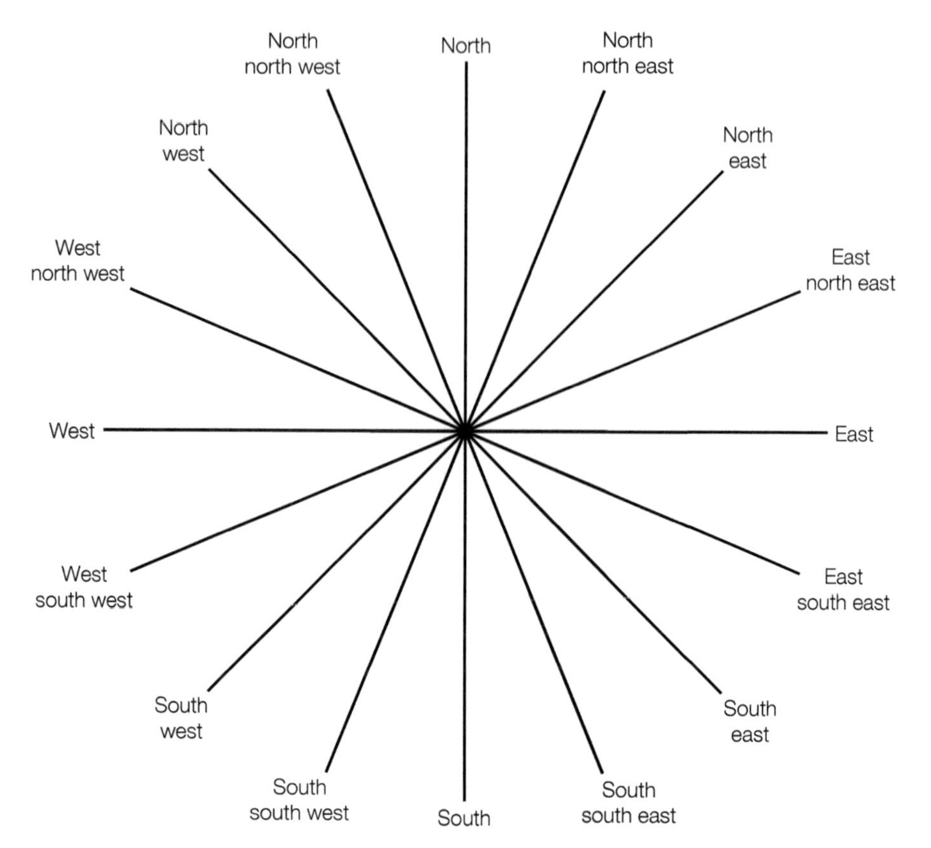

▲ **Fig. 28.5** *You are expected to know the 16 points of the compass*

360° bearings

This is an accurate way of showing direction clockwise from grid north. Use a protractor exactly over the point you wish to measure from and aligned north to south along the grid lines as shown below.

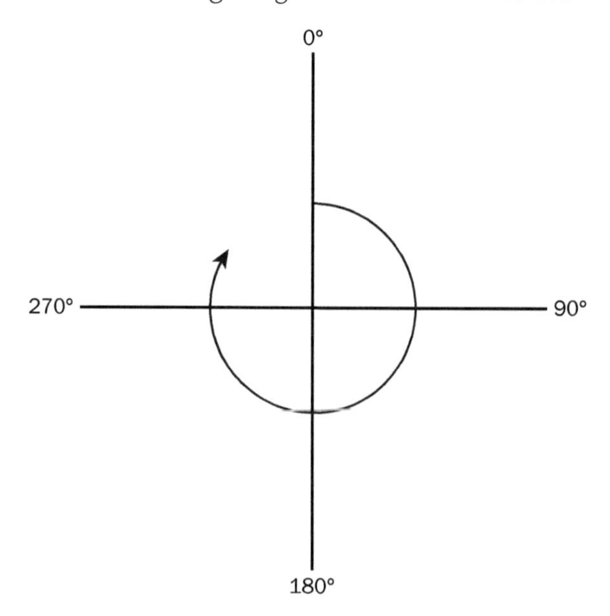

▲ **Fig. 28.6** *How to measure a 360° bearing*

Spot heights and trigonometrical points

A **spot height** is a dot on the map with a number beside it, which shows the number of metres that the point indicated is above sea level. Sometimes the spot height is combined with a **trigonometrical point** (station).

Contours

A **contour** on a map is a line that joins places of equal height above sea level. The difference in height between the contours (sometimes called the contour interval) varies, but is often 10 metres or 20 metres. Important contours such as 100 metres, 200 metres, 300 metres, etc. are often shown by a bold line.

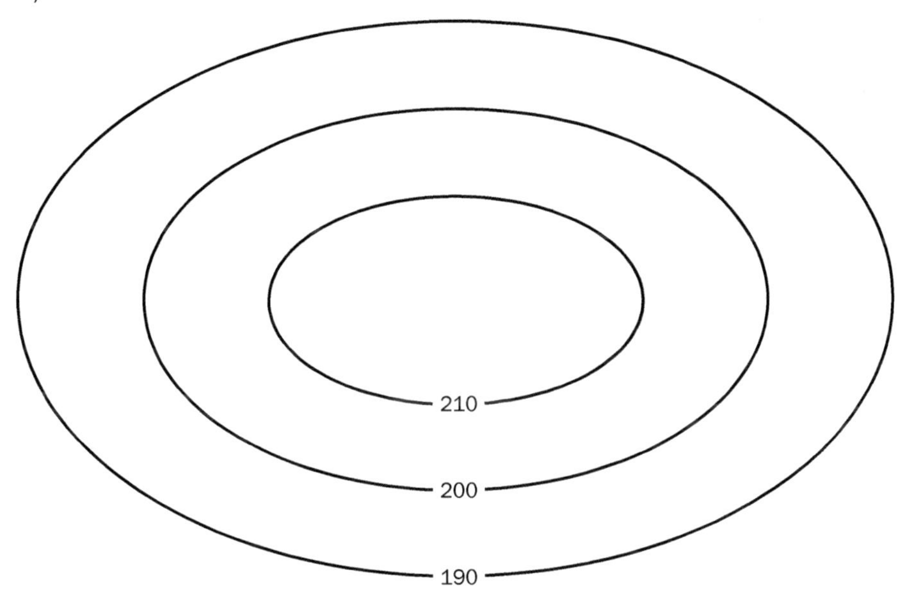

▲ **Fig. 28.7** *The contour pattern of an oval shaped hill; the summit is over 210 metres but doesn't reach 220 metres*

Exam tip

Exam questions often ask candidates to complete a relief cross-section and to add labels to it. Make sure that you have practised these skills. See Chapter 30 for more details.

The questions in this chapter are all worth one mark. You can check your answers with the mark scheme.

Sample question

1.

(a) How big are the grid squares on a 1:25 000 map? **[1 mark]**

(b) How big are the grid squares on a 1:50 000 map? **[1 mark]**

(c) Is an atlas map a large-scale map or a small-scale map? **[1 mark]**

(d) Is a survey map a large-scale map or a small-scale map? **[1 mark]**

See www.oxfordsecondary.com/esg-for-caie-igcse for the mark scheme for this question.

Analysis

✓ It is important to remember that there is always a scale line at the bottom of the map to help.

✓ Grid squares on survey maps used in the exam always have an area of 1 km^2.

Sample question

2. A road is shown crossing the map.

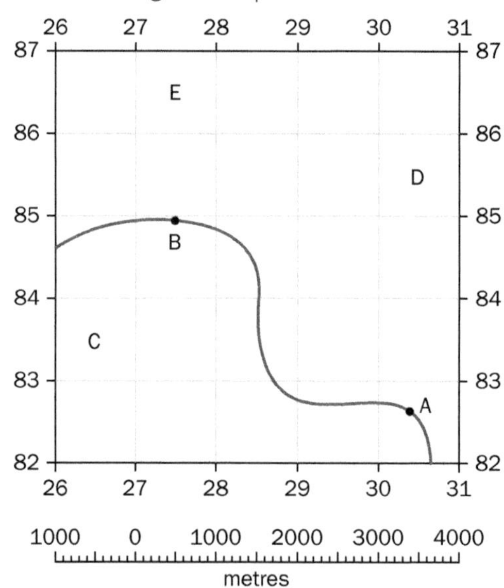

What is the distance between points A and B:

(a) in a straight line? **[1 mark]**

(b) along the road? **[1 mark]**

See www.oxfordsecondary.com/esg-for-caie-igcse for the mark scheme for this question.

Analysis

✓ Remember the comment earlier in this chapter about how to measure distances.

✓ To measure along the road, use the edge of a sheet of paper, breaking it down into short, straight sections.

✓ Give the correct units, otherwise you might not get the mark.

✓ The examiners will always allow some tolerance, so your answer doesn't have to be exactly correct.

Raise your grade

Sample question

3. On the same map, give the four-figure grid references for the following grid squares.

(a) C **[1 mark]**

(b) D **[1 mark]**

(c) E **[1 mark]**

See www.oxfordsecondary.com/esg-for-caie-igcse for the mark scheme for this question.

Analysis

✓ Remember that eastings come before the northings and to give the bottom left-hand corner of the square.

Sample question

4. Give the six-figure grid references for the following points:

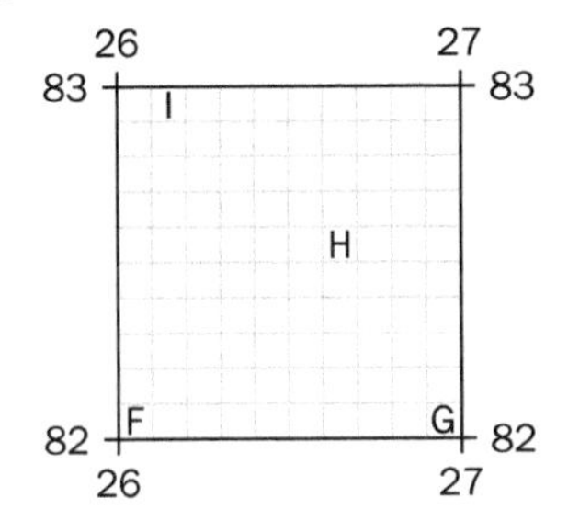

(a) F **[1 mark]**

(b) G **[1 mark]**

(c) H **[1 mark]**

(d) I **[1 mark]**

See www.oxfordsecondary.com/esg-for-caie-igcse for the mark scheme for this question.

Analysis

✓ All the answers are in the same square therefore all are 26_82_.

✓ Remember the advice earlier in this chapter about giving the third and sixth figures.

Sample question

5. Look at the map for Question 2.

(a) What is the compass direction from point A to point B? **[1 mark]**

(b) What is the compass direction from point B to point A? **[1 mark]**

(c) What is the 360° bearing from point A to point B? **[1 mark]**

(d) What is the 360° bearing from point B to point A? **[1 mark]**

See www.oxfordsecondary.com/esg-for-caie-igcse for the mark scheme for this question.

Analysis

✓ Examiners would allow a range of answers within a reasonable tolerance, so your answers don't have to be exactly correct.

Physical and human features on survey maps

Key ideas

- **Physical features on survey maps include relief and drainage.**
- **Human features on survey maps include settlement, communications and land use.**

Common error

The term 'relief' is often misunderstood. Learn the definition given here.

Relief features

Slopes

The closeness of the contours shows the steepness of the slope. Closely spaced contours indicate a steep slope; widely spaced contours show a gentle slope. The absence of contours may indicate flat land. Cliffs are shown by a separate symbol.

Uplands and lowlands

The contour heights and spot heights on the map show the height above sea level and can be used to show the higher and lower areas on a map. There is no precise definition as to how high or low an area has to be to be classified as highland or lowland. In some areas of the world entire countries are high above sea level.

Valleys

Small valleys without a flat floor are shown on maps by a V-shape in the contours. The V always points to high ground. There may or may not be a river in the centre.

Exam tip

When describing **relief** on maps, try to answer the following questions:

- Is the area highland or lowland?
- What is the average height and the height of the highest point?
- Are there areas which are steep, gentle, flat or cliffs?
- Are there any features such as valleys, **floodplains**, **plateaux**, **ridges**, **spurs** or **scarps**?

Cross-sections

A cross-section is a type of diagram often used in examination questions. It is as if a giant has sliced the landscape vertically along a line and pulled it apart.

Cross-sections are drawn to scale. The horizontal scale is generally the same as the map scale, but the vertical scale is made bigger (vertical exaggeration), so that features such as hills and valleys show up better. The position of features on the ground surface can be shown with labelled arrows.

There is an activity on cross-sections at the end of the chapter.

Exam tip

Exam questions may ask you to complete a cross-section and label features on it. Make sure that you know how to do this. Try the activity at the end of this chapter.

Drainage features

Drainage includes rivers and streams and their features, lakes and ponds. Marsh may be considered to be a feature of the drainage, i.e. poorly drained land, or it may also be considered a feature of the vegetation, where the plants are adapted to these conditions. Drainage may also include features produced by human activity, such as drainage channels, dams or reservoirs.

Areas of high **drainage density** have lots of surface water and areas of low drainage density have very little, often due to permeable rocks, such as limestone, which cause the water to seep underground.

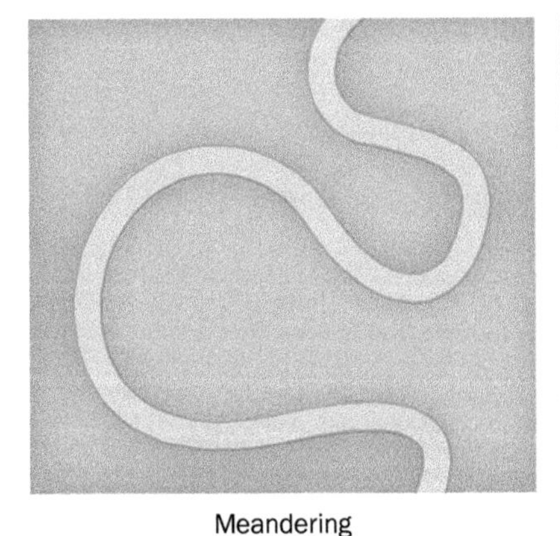

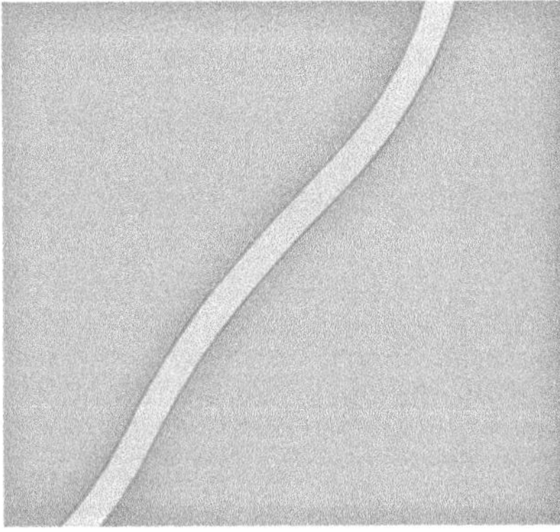

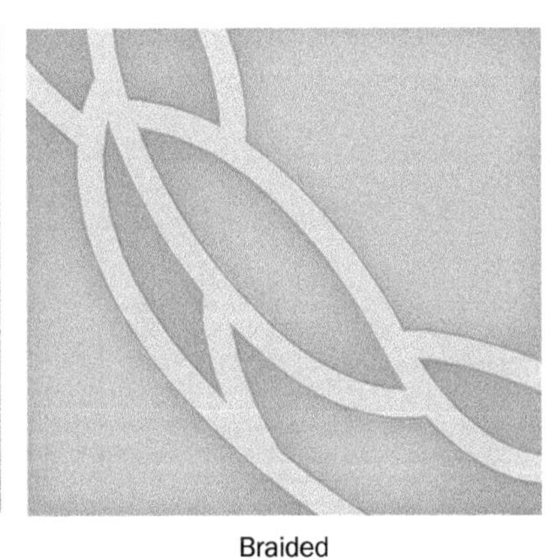

▲ **Fig. 29.1** *Channel shape*

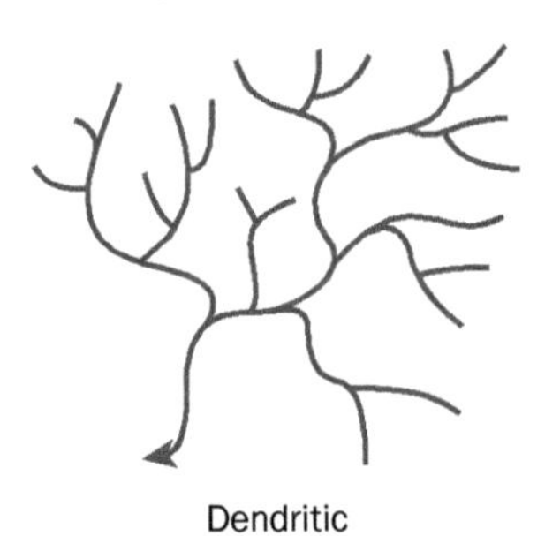

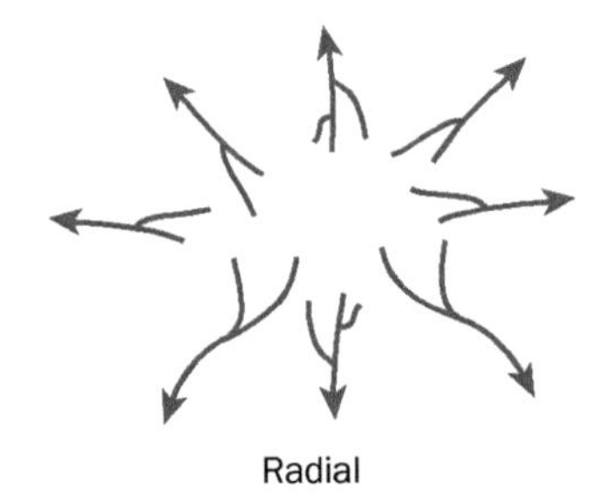

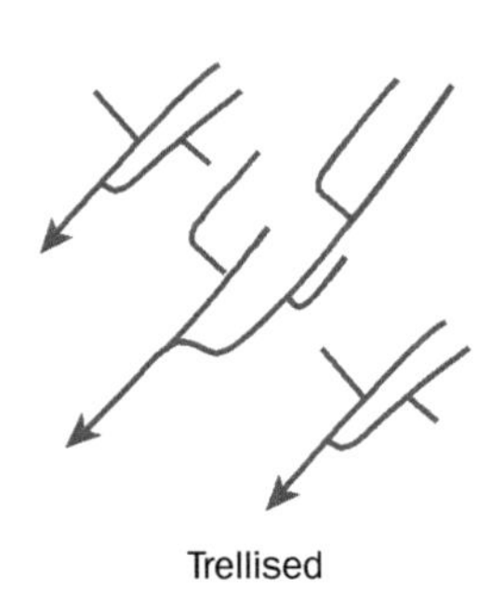

▲ **Fig. 29.2** *Stream (drainage) pattern*

Exam tip

When describing drainage on maps, try to answer the following questions.

- Is there a main river, what is its width and are there small **tributaries**?
- What is the flow direction?
- Is the drainage density high or low?
- What is the channel shape?
- What is the drainage pattern?
- Are there features such as dams, waterfalls, rapids?

Common error

Candidates often think that rivers flow inland from the sea and split into tributaries. This is completely wrong.

Common error

Candidates often confuse *physical* features with *human* features. Make sure that you know the difference.

Coastlines

When describing **physical features** of coasts on maps try to answer the following. Are there features such as:

- bays or headlands
- estuaries or river mouths
- beaches
- coral reefs
- cliffs
- wave-cut platforms
- stacks
- marshes or swamps?

Distribution, density and location of settlements

These are usually affected by factors such as:

- availability of communications
- accessibility of points, such as road junctions (route centres) and bridge points
- availability of cultivated land
- avoiding steep slopes
- avoiding land which is liable to flooding and may also be affected by pests and disease.

Settlement patterns: nucleated, dispersed, linear

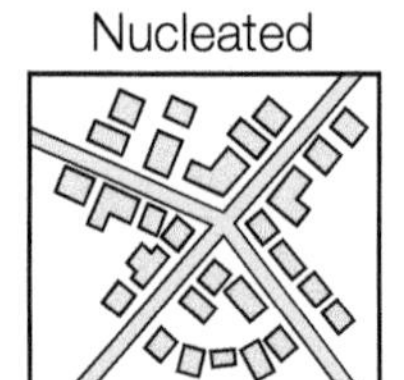

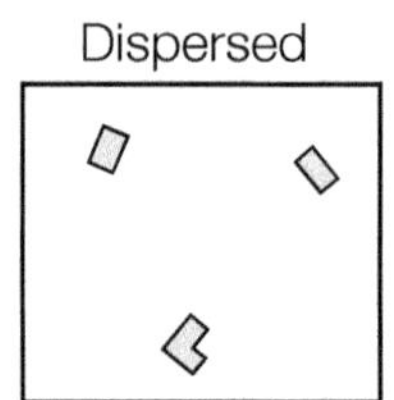

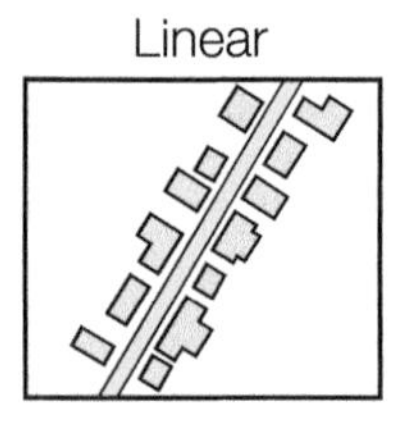

▲ **Fig. 29.3** *Nucleated, dispersed and linear settlements*

Exam tip

Make sure that you know the difference between a cutting and an embankment and the symbols used for them.

Communications

Communications on maps are generally the different types of roads, tracks and railways and, occasionally, ports and airports/airstrips. Care should always be taken to read the map key carefully to identify these features correctly.

Links with physical and human features

Roads usually try to follow gentle slopes. They try to avoid steep slopes and areas which are liable to flood. For this reason, they often follow valleys, at the bottom of the valley sides and avoid floodplains. When steep slopes are encountered, roads may zigzag and have hairpin bends to make the gradient more gentle.

Railways need very gentle gradients. They often have cuttings or tunnels through hills or cross lowland areas on embankments.

Land use

The land use symbols used on maps vary greatly from country to country. Typically, they show natural vegetation, types of cultivation and settlement.

Sample question

1. Describe the relief of the area shown on the map. **[4 marks]**

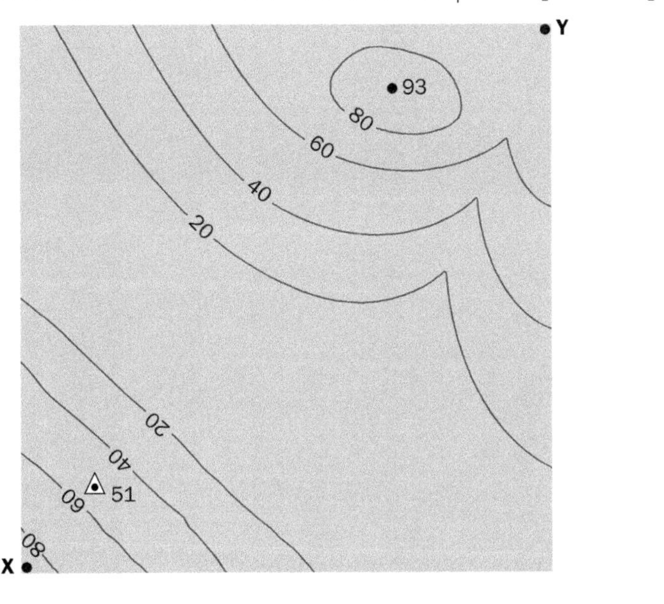

Analysis

✓ Tips on how to answer questions about relief have already been given in this chapter.

✓ This is a frequent question on exam papers.

Student answer

The area is fairly low with the highest point at 93 and the lowest at 20. It shows a valley with a flat floodplain. There is a small spur in the north-east of the map. [3 marks]

See www.oxfordsecondary.com/esg-for-caie-igcse for the mark scheme for this question.

Examiner feedback

Although there are some good points here, the candidate has not given the units for the heights of 93 and 20 metres, therefore doesn't score any marks.

Three marks are awarded for the mention of lowland, valley and floodplain. The valley floor may be below 20 metres. The V-shape in the contours in the north-east of the map points to the high ground. Therefore, this is a small, V-shaped, tributary valley and not a spur.

Sample question

2. Describe the drainage of the area shown on the map. **[5 marks]**

Analysis

✓ Tips on how to answer questions about relief have already been given in this chapter.

✓ This is a frequent question on exam papers.

Student answer

See www.oxfordsecondary.com/esg-for-caie-igcse for the mark scheme for this question.

The main river has meanders and one island. It is joined by four tributaries. The main tributary has a dendritic pattern. The drainage density is high in the north-east, but low in the south-west. One tributary has a dam. There is an area of marsh in the north-west. [5 marks]

Examiner feedback

This is a good answer. Actually more than five points have been made. Can you see them?

Sample question

3. The cross-section has been drawn from X to Y across the map used for Question 1.

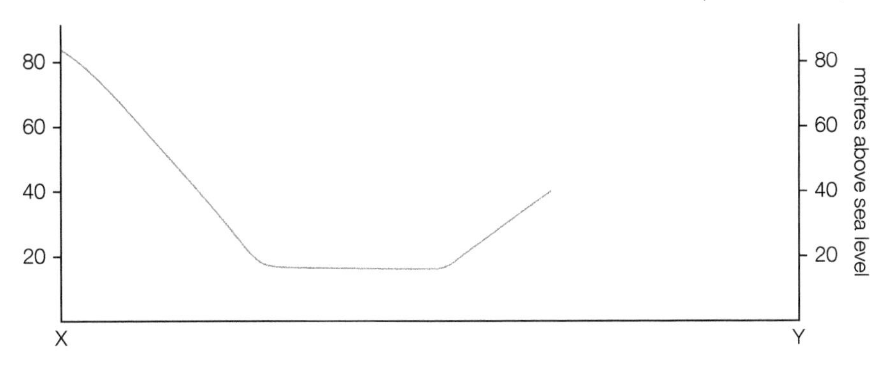

On the cross-section:

(a) Use labelled arrows to show the positions of the floodplain and a trigonometrical point at 51 metres. **[2 marks]**

(b) Complete the cross-section. **[2 marks]**

Analysis

✓ Make sure your labelled arrows end exactly on the cross-section.

Student answer

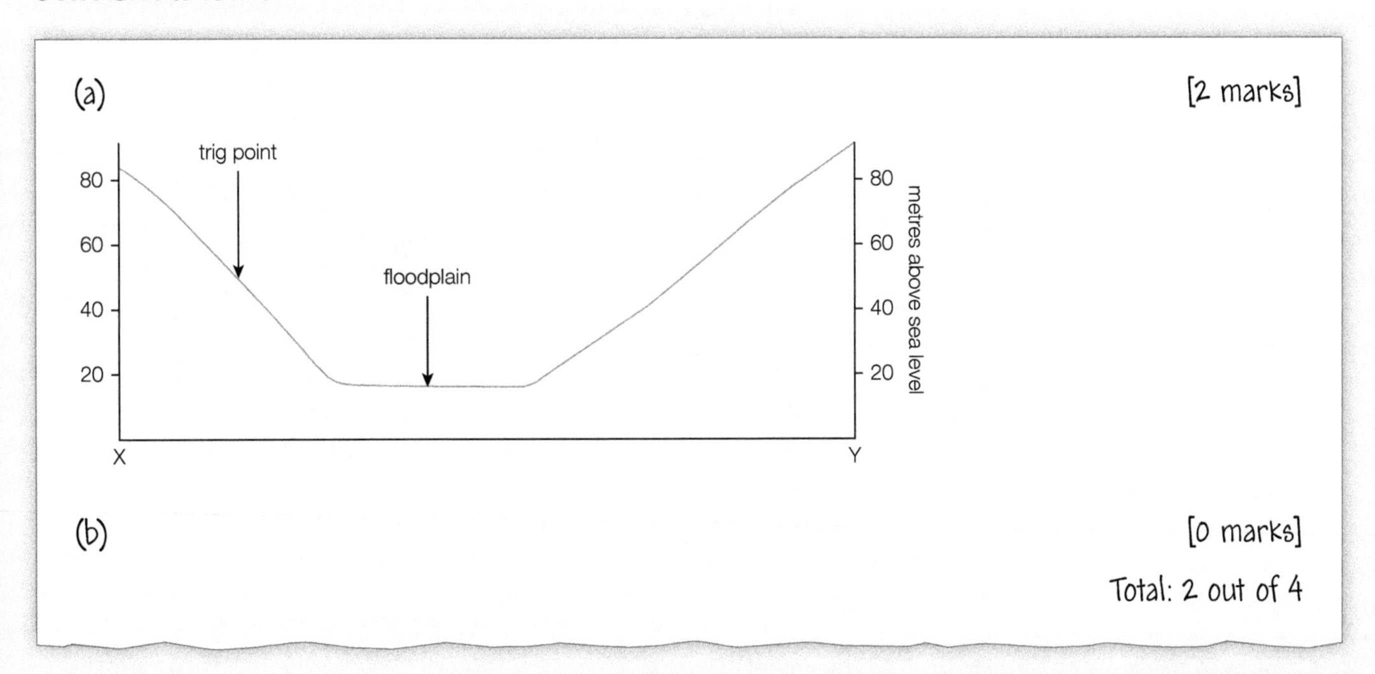

See www.oxfordsecondary.com/esg-for-caie-igcse for the mark scheme for this question.

Examiner feedback

The positions of the trigonometrical station and the floodplain have been labelled clearly and accurately. However, the cross-section has not been completed correctly. The land should rise to almost 80 m in the north-east, but after the summit it drops again to below 80 m, as shown below.

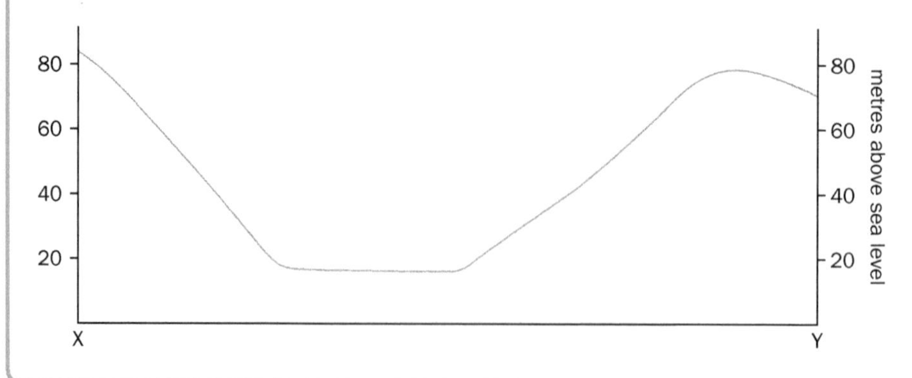

Sample question

4. Look at the settlement shown on the map.

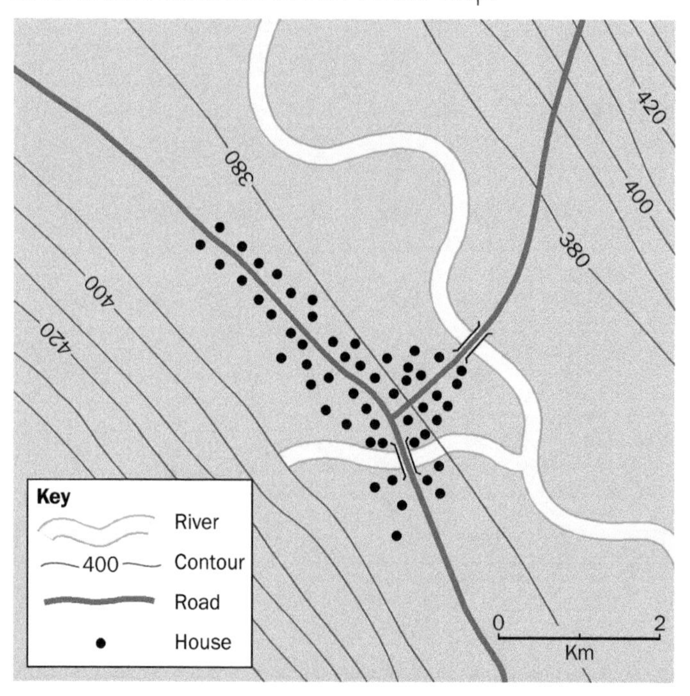

How has the location of the settlement been affected by:

(a) relief **[3 marks]**

(b) drainage and water supply **[2 marks]**

(c) transport? **[2 marks]**

Analysis

✓ Tips on how to answer questions like this are given earlier in this chapter.

Student answer

(a) The settlement is on gently sloping land at the foot of the steep valley side. The slope is facing north-east. [2 marks]

(b) It is on a river for a water supply, but only a small part of the settlement is on the flat valley floor which is liable to flood. [2 marks]

(c) It is located at a road junction, which makes it a very accessible point. [2 marks]

Total: 6 out of 7

See www.oxfordsecondary.com/esg-for-caie-igcse for the mark scheme for this question.

Examiner feedback

In part (a) two marks are awarded for *gentle slopes* and *north-east facing*. Can you see where two marks were awarded in both parts (b) and (c)?

Key ideas

- These techniques are particularly important in Papers 2 and 4.
- Different types of graph are used for different purposes.

Photographs

You may be asked to describe the physical features (relief, drainage, vegetation) and the human features (settlement, agriculture, industry, transport) seen on a photograph. These are described on pages 322 and 323 of *Complete Geography for Cambridge IGCSE and O Level.*

Common error

When answering photograph questions, concentrate on what can be seen in the photograph and avoid speculation.

Field sketches

In examination questions, field sketches are sometimes used with photographs.

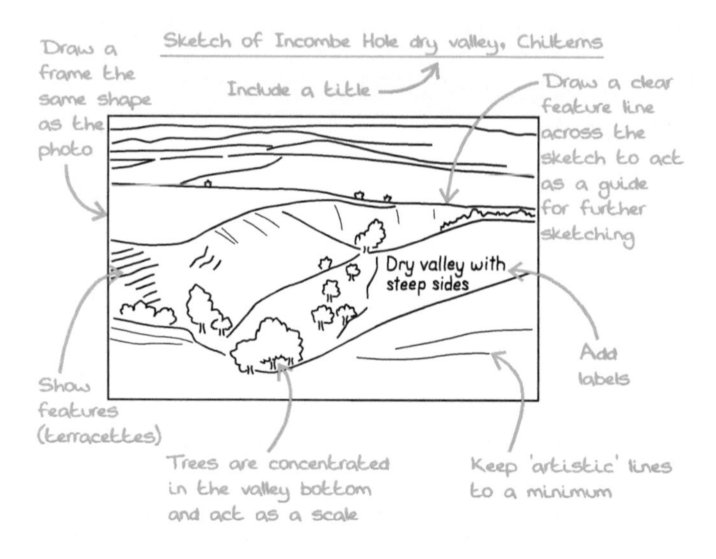

▲ **Fig. 30.1** *Notice how this field sketch has been clearly labelled*

Exam tip

Remember to label your sketches.

Data tables

Tables of data are often used in examination questions. The following table, which shows the number of births and deaths (thousands) in the UK, is an example. From 2021, the figures are projected rather than actual.

Year	1951	1961	1971	1981	1991	2001	2011	2021	2031	2041	2051
Births	790	940	900	740	790	670	780	780	770	830	840
Deaths	610	630	650	660	640	600	560	560	630	720	760

▲ **Fig. 30.2** *Can you pick out the trends in this data?*

Exam tip

You should be able to look at the data and identify any patterns or trends. Do not simply list individual years' data.

Graph types

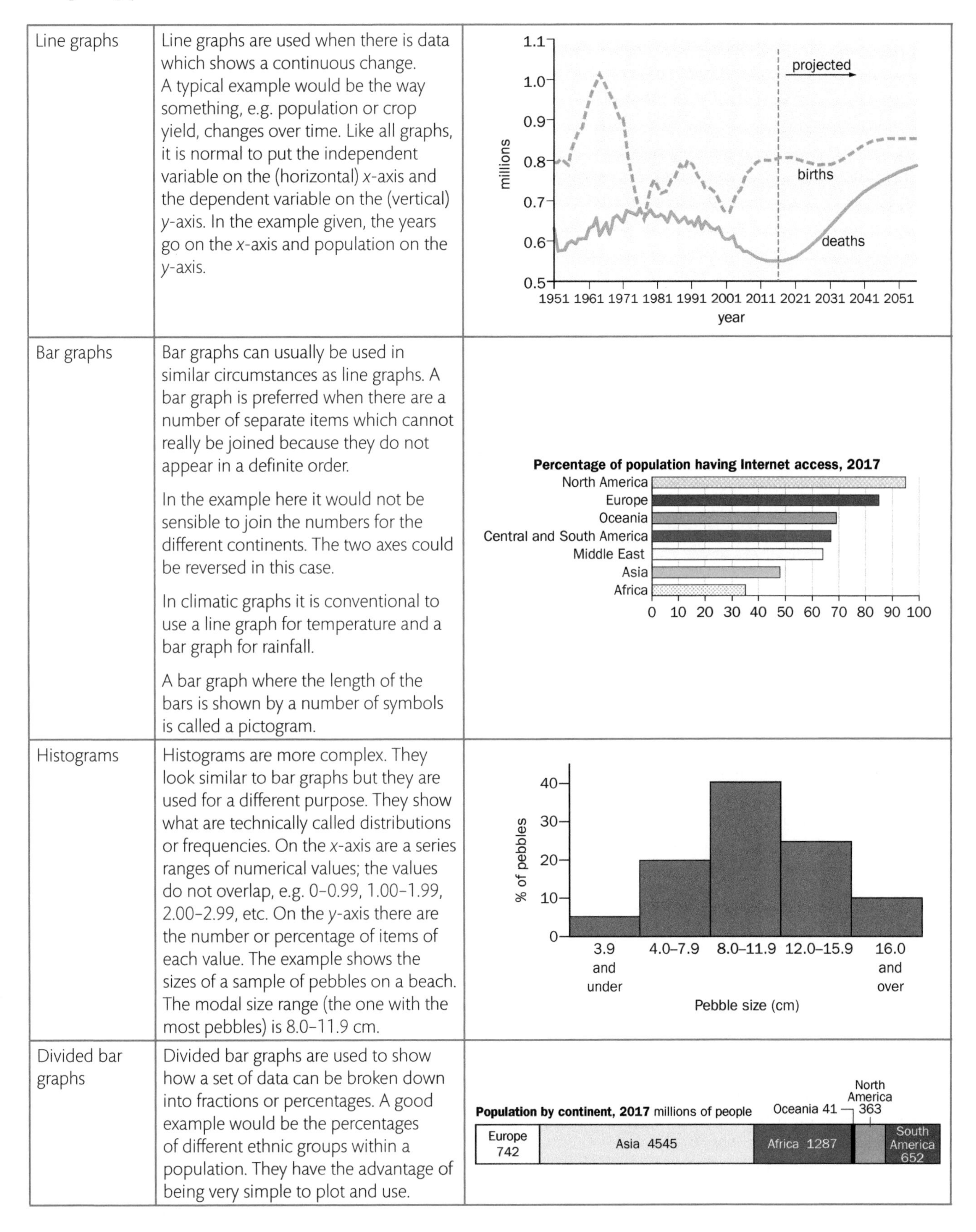

Graph type	Description
Line graphs	Line graphs are used when there is data which shows a continuous change. A typical example would be the way something, e.g. population or crop yield, changes over time. Like all graphs, it is normal to put the independent variable on the (horizontal) *x*-axis and the dependent variable on the (vertical) *y*-axis. In the example given, the years go on the *x*-axis and population on the *y*-axis.
Bar graphs	Bar graphs can usually be used in similar circumstances as line graphs. A bar graph is preferred when there are a number of separate items which cannot really be joined because they do not appear in a definite order. In the example here it would not be sensible to join the numbers for the different continents. The two axes could be reversed in this case. In climatic graphs it is conventional to use a line graph for temperature and a bar graph for rainfall. A bar graph where the length of the bars is shown by a number of symbols is called a pictogram.
Histograms	Histograms are more complex. They look similar to bar graphs but they are used for a different purpose. They show what are technically called distributions or frequencies. On the *x*-axis are a series ranges of numerical values; the values do not overlap, e.g. 0–0.99, 1.00–1.99, 2.00–2.99, etc. On the *y*-axis there are the number or percentage of items of each value. The example shows the sizes of a sample of pebbles on a beach. The modal size range (the one with the most pebbles) is 8.0–11.9 cm.
Divided bar graphs	Divided bar graphs are used to show how a set of data can be broken down into fractions or percentages. A good example would be the percentages of different ethnic groups within a population. They have the advantage of being very simple to plot and use.

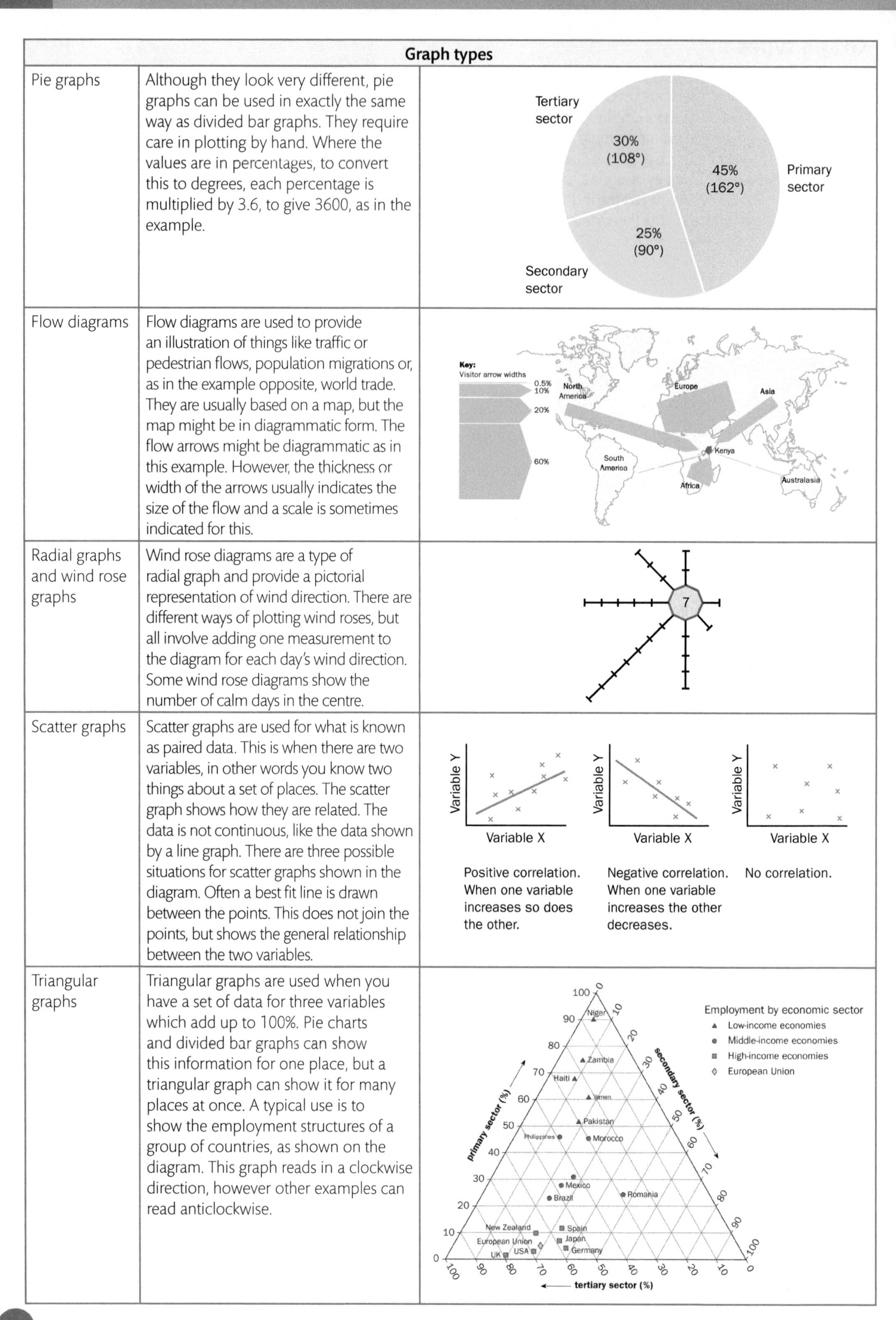

Graph types	
Pie graphs	Although they look very different, pie graphs can be used in exactly the same way as divided bar graphs. They require care in plotting by hand. Where the values are in percentages, to convert this to degrees, each percentage is multiplied by 3.6, to give 3600, as in the example.
Flow diagrams	Flow diagrams are used to provide an illustration of things like traffic or pedestrian flows, population migrations or, as in the example opposite, world trade. They are usually based on a map, but the map might be in diagrammatic form. The flow arrows might be diagrammatic as in this example. However, the thickness or width of the arrows usually indicates the size of the flow and a scale is sometimes indicated for this.
Radial graphs and wind rose graphs	Wind rose diagrams are a type of radial graph and provide a pictorial representation of wind direction. There are different ways of plotting wind roses, but all involve adding one measurement to the diagram for each day's wind direction. Some wind rose diagrams show the number of calm days in the centre.
Scatter graphs	Scatter graphs are used for what is known as paired data. This is when there are two variables, in other words you know two things about a set of places. The scatter graph shows how they are related. The data is not continuous, like the data shown by a line graph. There are three possible situations for scatter graphs shown in the diagram. Often a best fit line is drawn between the points. This does not join the points, but shows the general relationship between the two variables.
Triangular graphs	Triangular graphs are used when you have a set of data for three variables which add up to 100%. Pie charts and divided bar graphs can show this information for one place, but a triangular graph can show it for many places at once. A typical use is to show the employment structures of a group of countries, as shown on the diagram. This graph reads in a clockwise direction, however other examples can read anticlockwise.

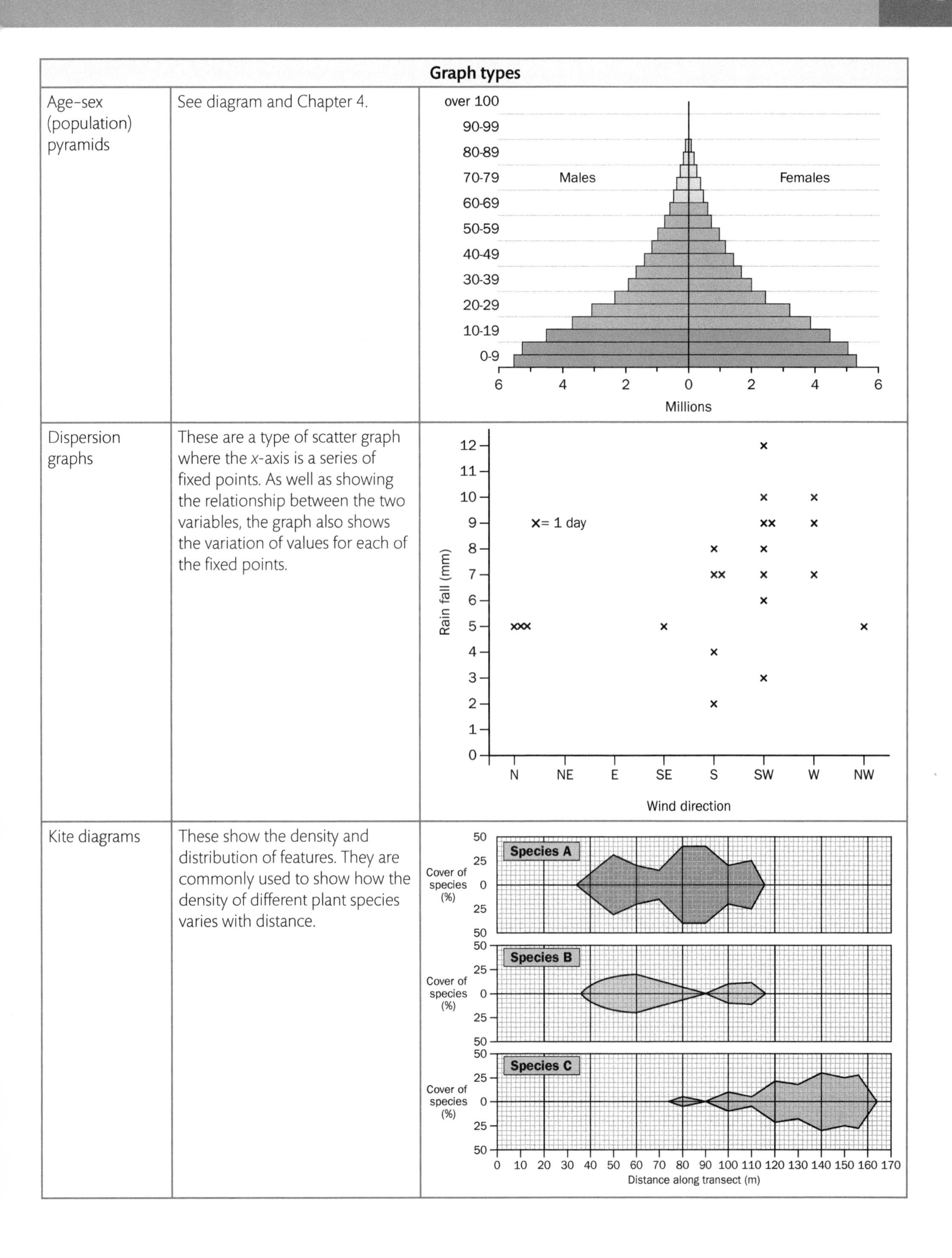

Graph types		
Age–sex (population) pyramids	See diagram and Chapter 4.	
Dispersion graphs	These are a type of scatter graph where the *x*-axis is a series of fixed points. As well as showing the relationship between the two variables, the graph also shows the variation of values for each of the fixed points.	
Kite diagrams	These show the density and distribution of features. They are commonly used to show how the density of different plant species varies with distance.	

Sample question

1. Study the survey map and answer the questions which follow.

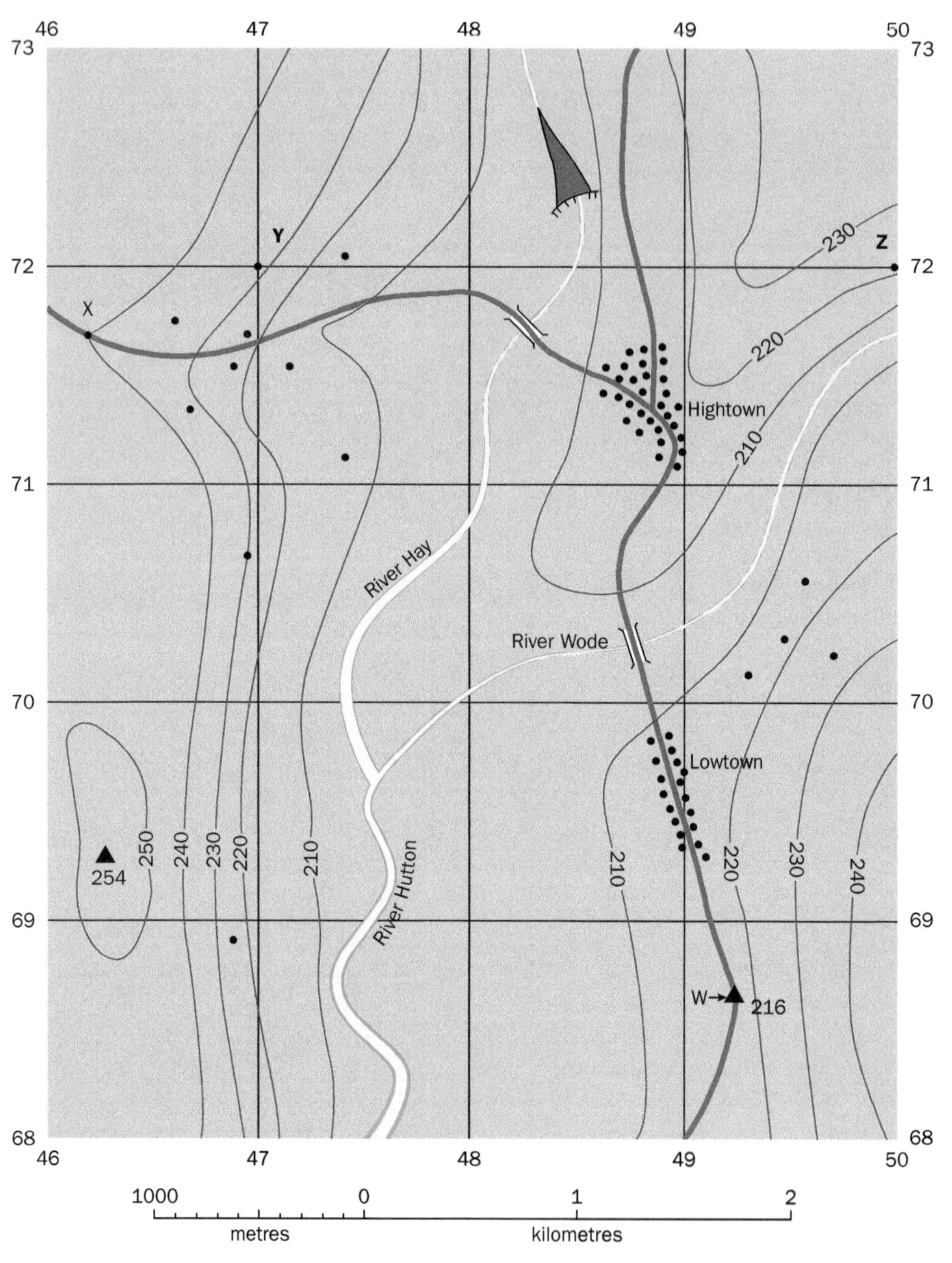

(a) (i) Give the four-figure grid reference of the village of Hightown. **[1 mark]**

(ii) Give the six-figure grid reference of the trigonometrical station on the road at point W. **[1 mark]**

(b) Measure the distance between points W and X:

(i) in a straight line; **[1 mark]**

(ii) along the road. Give your answers in metres. **[1 mark]**

(c) (i) What is the compass direction from point X to point W? **[1 mark]**

(ii) What is the 360° bearing from point X to point W? **[1 mark]**

Analysis

✓ This final question is to test all of your skills for answering Question 1 on Paper 2.

Student answer

(a) (i) 4871 [1 mark]

(ii) 491686 [0 marks]

(b) (i) 4300 [0 marks]

(ii) 5750 [0 marks]

(c) (i) South-east [1 mark]

(ii) 138° [1 mark]

Total: 3 out of 6

Examiner feedback

In part (a), the four-figure reference is correct. The six-figure reference is in the correct square, but the third figure should be 2.

In part (b), both the measurements are correct, but the correct units (metres) should be stated.

Both the answers to part (c) are correct.

See www.oxfordsecondary.com/esg-for-caie-igcse for the mark scheme for this question.

Sample question

(d) The following cross-section has been drawn along northing 720 from point Y at 470720 and point Z at 500720. The cross-section has not been completed.

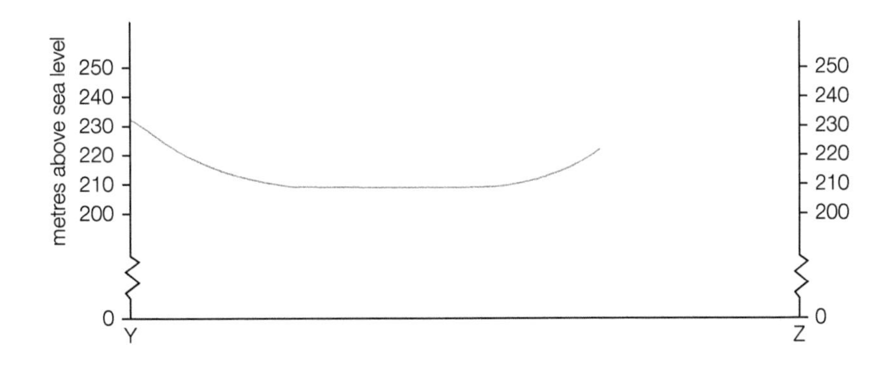

(i) On the cross-section, use labelled arrows to show the position of:

- the River Hay
- the road
- the river floodplain
- the slope on the west side of the valley. **[4 marks]**

(ii) Complete the cross-section. **[2 marks]**

Student answer

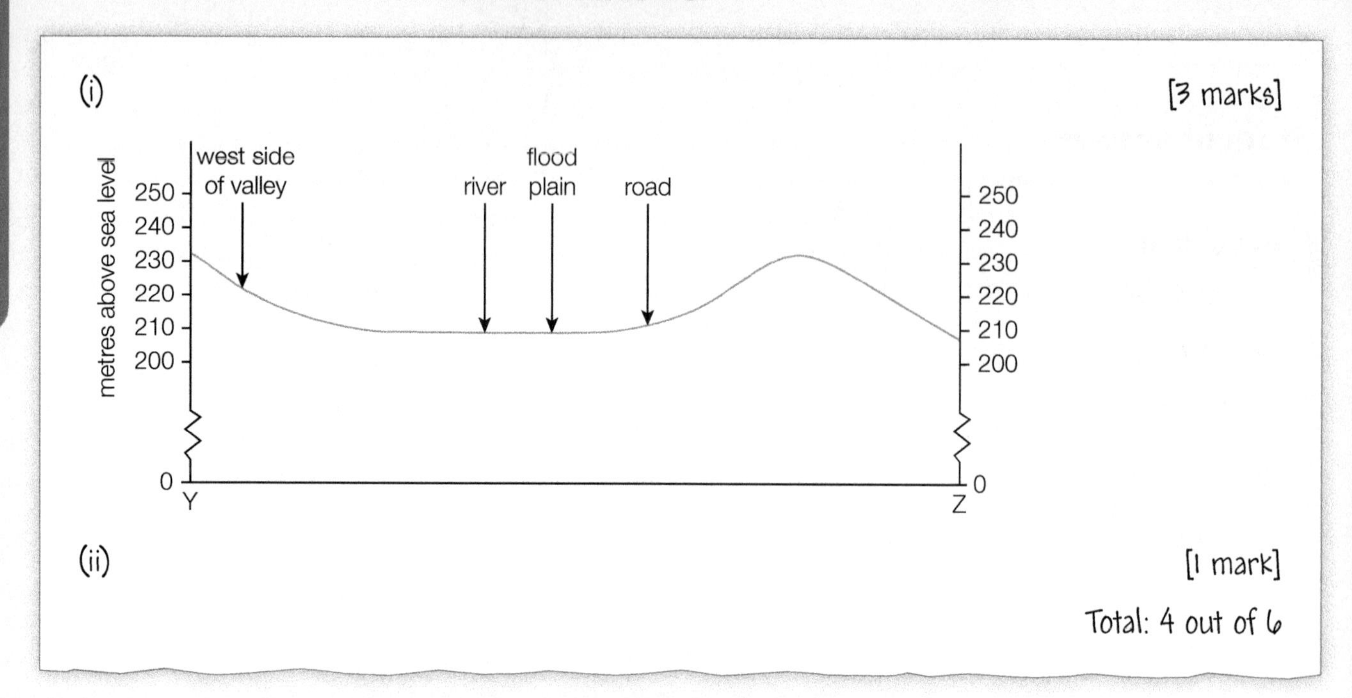

See www.oxfordsecondary.com/esg-for-caie-igcse for the mark scheme for this question.

Examiner feedback

The west side of the valley, road and floodplain are labelled correctly, but the river is not accurate. The correct method of labelling has been used. When plotting the cross-section, the rise in the ground to over 230 metres is correct, but the ground should then drop to almost 220 metres at Z.

Sample question

(e) Describe the relief of the area south of northing 700. **[5 marks]**

(f) Describe the drainage in the area of the map. **[4 marks]**

(g) Houses are shown on the map by circles. Describe the settlement pattern on the map. **[5 marks]**

(h) Describe how the roads in the area of the map are affected by the relief. **[4 marks]**

Student answer

(e) The area is a valley with the River Hutton and the village of Howtown. There is one road to provide relief. [1 mark]

(f) The river meanders and is wider in the south. There is a dam in the north. The river flows north and splits into tributaries. [3 marks]

(g) Hightown is a nucleated settlement and Lowtown is a linear settlement. The rest of the settlement is dispersed. [3 marks]

(h) The roads are on gentle slopes at the foot of the valley sides. They avoid the floodplain and the steeper slopes. Occasionally a road has to cross the floodplain. At Hightown the road is on a spur and in the north west it climbs a slope by a small side valley. [4 marks]

Total: 11 out of 18

Examiner feedback

In part (e), the candidate has misunderstood the term *relief*. The only relevant point is about the valley. Points which could have been included are the flat floodplain, about 1.5 km wide, the steeper valley sides with the west side steeper than the east and the high point of 254 metres.

In part (f), rivers do not split into tributaries. The river flows south and is joined by a tributary, the River Wode.

The answer to part (g) is correct, however more marks could have been scored by noting the linear settlement being *along the road*, and the nucleated settlement being *at a road junction*. Also, the settlement is generally on gentle slopes. It avoids the floodplain.

Part (h) is an excellent answer.

See www.oxfordsecondary.com/esg-for-caie-igcse for the mark scheme for this question.

Coursework skills

Many of the graphical, statistical, analytical and investigation skills used in coursework can also be very useful in general life.

Chapters 31 to 33 are designed to help you understand the stages of a geographical investigation, including planning (and formulating a hypothesis), collecting data, presenting data in different forms, analyzing and interpreting the data, drawing conclusions based on the results, and evaluating the success of the investigation.

You will be taking either Paper 3 (submitting your own piece of coursework) or Paper 4 (the alternative to coursework paper). The questions on Paper 4 are each worth 30 marks and go through the stages of a geographical investigation described above.

Coursework: Planning investigations

Key ideas

- Careful planning is essential to good coursework enquiries.
- The objective of the study should involve a problem, issue or question that arises from studying content in the syllabus.
- A hypothesis must be designed that will test the problem, issue or question.
- Decisions have to be made about what data will be relevant to the study.
- Methods of accurately collecting the data must be thoroughly planned.

Choosing the topic

The topic to be investigated should be:

- a **hypothesis** or question likely to succeed
- able to be completed in the time available by the number of people in the group
- able to be done in an accessible and safe location
- able to be done using equipment available
- likely to succeed.

Formulating a hypothesis

To see if an expected relationship exists, a possible format is: 'X increases as Y decreases' or 'There is a negative relationship between X and Y'. Another way is to test a statement, such as, 'Larger settlements have more comparison shops'.

Health and safety considerations

- Be aware of the dangers of working near and in rivers, especially on outer bends of meanders where the water is faster flowing and deeper and the bank may overhang. Remember riverbanks can be steep and slippery.
- Be careful near the sea, which can have freak waves as well as strong undercurrents out to sea. Also wooden groynes can trap feet in gaps and cliffs can suddenly collapse, especially after heavy rain.
- Tide tables should be consulted before working on a beach or walking under cliffs away from the access to the beach.
- Take safety precautions, which may include never working alone and having a mobile phone with its battery fully charged with home and school contact numbers stored in it. Beforehand, make sure someone at home knows where you will be, who with and your expected time schedule.
- Wear strong footwear and suitable clothing, including a life vest if appropriate.
- Consider the need to wear insect repellent and sunblock.
- Questionnaires should be conducted in pairs in safe locations, even in towns, where congested pavements should be avoided.

Common error

When planning to research a relationship, remember that all other variables, such as the time of the survey, must be kept the same for all the research so that they cannot influence the results.

Common error

Remember to make a checklist of what is needed in the field.

Common error

A pilot survey is not one undertaken from a plane!

Choosing how to investigate your hypothesis or statement

You may need to use:

- large-scale maps to locate points where the investigation will be done
- recording sheets designed to fit your investigation; you may also record on **field** sketches, maps and photographs
- **questionnaires** to interview people
- an appropriate type of sampling to decide where or who to **sample**
- equipment needed to obtain recordings for your investigation
- pairs or larger groups to obtain recordings at different locations simultaneously
- **secondary data**
- a **pilot study (practice survey).**

Questionnaires

A questionnaire is used to find out the attitudes of the public or gain information from them. It should have:

- the headings needed for a recording sheet – title, date, time, location, name of student, with spaces for their completion
- an introductory sentence about the purpose of the questionnaire
- a first question which ensures that the person will be relevant to the study, for example someone local to the area, not a visitor
- a short series of essential closed questions with options and spaces for tick-box answers, starting with questions that put the interviewee into categories as necessary to the study, such as age range, gender
- questions that are simply and clearly expressed and do not use terminology the public may not know.

It is necessary to consider the following when using a questionnaire:

- what type of sampling to use
- what form the pilot study should take
- where to do the survey.

Sampling

- A sample aims to investigate the smallest number that would be large enough to be *truly representative of the whole population.*
- If differences are significant to the investigation, samples can be divided into subgroups with different characteristics, for example age groups or types of vehicle.
- To be a fair test, the sample must be chosen without bias; this means that *every individual in the population must have an equal chance of being included in the investigation.* The investigator must not choose which people should be asked to answer questions.
- The larger the sample, the more reliable the results are likely to be. A sample of 30 is usually sufficient when a relationship is being investigated.
- It is important that the most appropriate sampling method is used.
- **Random sampling** removes bias from an investigation. Random number tables are used to decide what or where to sample. Part of a random number table might look like this.

39	26	02	11	98	55
58	07	46	60	77	04
17	83	29	32	41	36
48	65	08	93	55	69

- The numbers can be read in any direction, as long as it is consistent.
- **Systematic sampling** ensures that the whole area or population under study is covered completely. It is used when a regular change is expected.
- **Stratified sampling** makes the sampling representative of the different subsets of the population. It is usual to balance genders and age groups. Stratified sampling has the important advantage that all parts of an area or sections of a population are included. Significant differences can be found between subgroups.
- Sampling can be of points, areas or points or areas along lines.

Common error

Remember that the results of sampling can never be described as 'accurate'. Use 'reliable' instead.

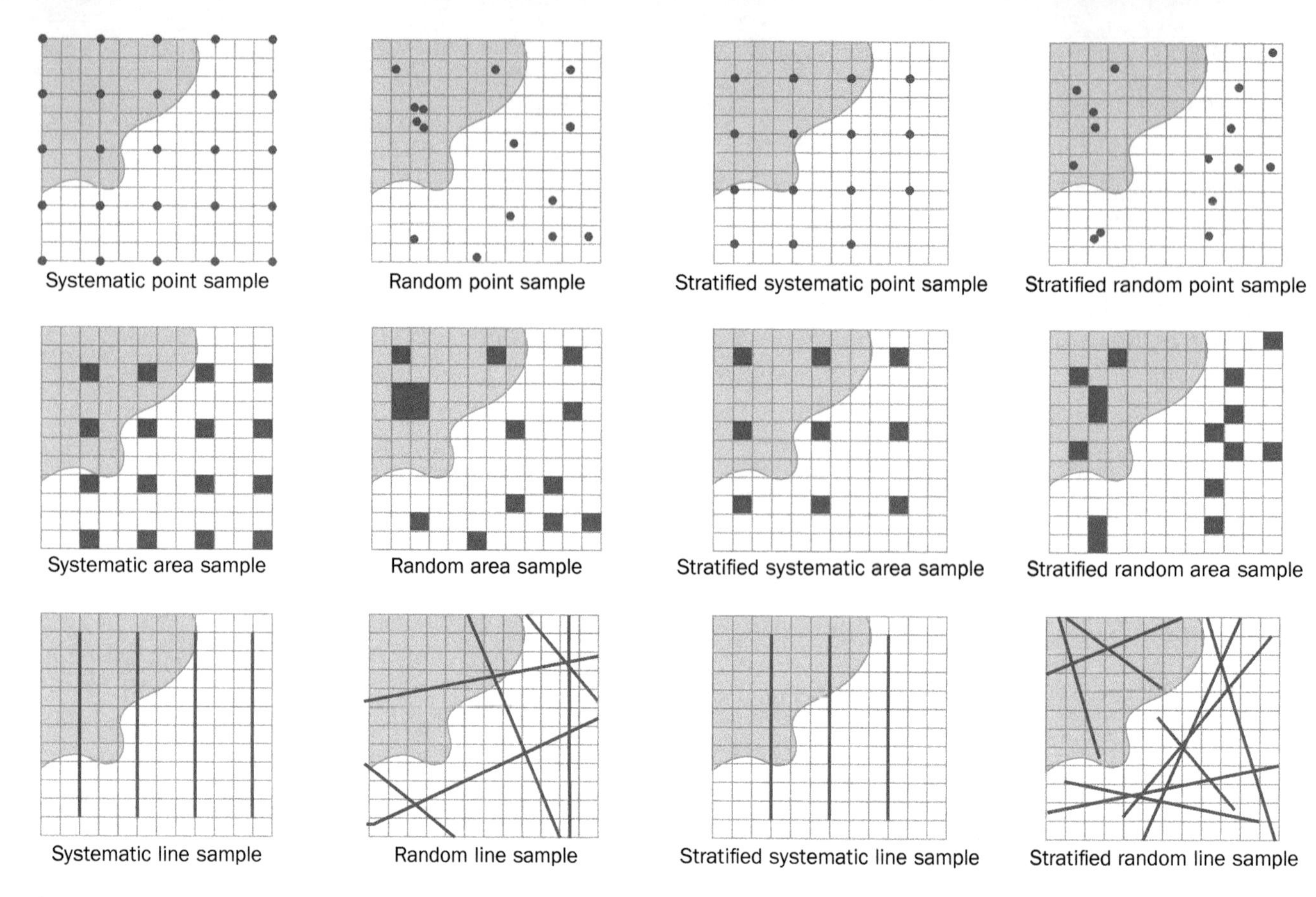

▲ **Fig. 31.1** *Examples of types of sampling*

Exam tip

The Alternative to Coursework examination tests your knowledge of how to make a successful geographical enquiry, so expect to be asked about any aspect of it, including all the topics in this chapter.

Recording sheets

The layout of these will depend on what observations are to be undertaken, but they need identification (e.g. time, date, location, name of recorder) and an appropriate number of rows and columns in which to record.

Student answers and examiner feedback

For chapters 31–33 you should consider how you would answer the questions worth 1 or 2 marks and check your answers with the mark scheme. The mark schemes can be found at www.oxfordsecondary.com/esg-for-caie-igcse.

For each question worth 3 marks or more in the following pages, a specimen answer is provided with examiner feedback.

These coursework chapters will cover a variety of the types of question you could be asked in the Alternative to Coursework examination and that you would need to consider if doing a coursework investigation.

Raise your grade

Sample question

1. (a) What are the reasons for, and advantages of, sampling? **[2 marks]**

   (b) Why is the choice of the sampling interval important? **[3 marks]**

   (c) Describe how you would sample if investigating changes in soil depth down a valley side. **[2 marks]**

   (d) You are sampling a representative sample of 30 farmers in an area in which 50% are arable farms, 30% are mixed farms and 20% are pastoral. How many of each type of farmer should you plan to include for a stratified sample? **[1 mark]**

   (e) Explain how you would choose which 30 of the 65 householders living in a village to visit, in order to find out their opinions about a proposed new housing development. Justify your choice of sampling method. **[3 marks]**

   (f) Explain why line and area sampling would not be appropriate to use for the sample required in part (e). **[2 marks]**

   Finally, check your knowledge of this important topic by completing the following task:

   (g) Draw two columns, one entitled 'Random sampling' and the other 'Systematic sampling'. Find opposites in the following descriptions and put them in the correct column at the same level. (Five of the eight are systematic, so complete them first.)

   - Can give a false idea by giving too few or too many counts of a feature if they are arranged in regular lines.
   - Not all features or people have an equal chance of being chosen.
   - Quicker and easier to do.
   - Ensures coverage of the population and prevents clusters being selected.
   - Takes longer in the field because of the irregular spread of points, areas or lines.
   - It is not always representative as it can miss small quantities of data and points can be clustered.
   - Without bias. **[4 marks]**

See www.oxfordsecondary.com/esg-for-caie-igcse for the mark scheme for this question.

Analysis

✓ The success of an investigation depends on the most appropriate sampling method being chosen.

✓ Generally, 30 results are considered sufficient to be a representative sample.

Student answer

1. (b) Student answer

A small sampling interval would result in the survey taking a lot of time, whereas a large interval could lead to some important features not being included. [2 marks]

Examiner feedback

The candidate includes the first two points in the mark scheme. Remember that the aim of sampling design is to have a representative sample of the population, so a sample interval that would be small enough to give at least 30 results should be used.

Student answer

1. (e)

Number each house on a map of the village. Read down random number tables and use the last two figures in each number to select the 30 houses.[2 marks]

Examiner feedback

The method is described clearly but the answer does not justify its choice. The third mark is for stating why random sampling is better than systematic sampling, which would introduce a bias.

Student answer

(g)	
Random sampling	Systematic sampling
Without bias (everyone has a chance of being selected)	Not all features or people have an equal chance of being chosen
Not always representative as it can miss small areas or quantities of data and points can be clustered	It ensures coverage of the population and prevents clusters being selected.
It can take longer to do because it can result in more movement between the irregular spread of points.	Quicker and easier to do.

Examiner feedback

Three differences have been included but the disadvantage that it can give a false idea if points are arranged in regular lines was omitted.

Sample question

2. (a) State precautions you should take to ensure your health and safety before you undertake any investigation in the field, whether in a rural area or a town. **[3 marks]**

(b) State additional health and safety considerations and precautions for work:

(i) on a beach **[1 mark]**

(ii) along a river **[1 mark]**

(iii) in a town. **[1 mark]**

See www.oxfordsecondary.com/esg-for-caie-igcse for the mark scheme for this question.

Analysis

✓ Part (a) states general precautions. In areas where there are other local dangers, such as dangerous wild animals, take other precautions.

✓ If the examiner considers these are marks that are too easily gained, only one of a similar pair might be credited, e.g. mobile phone or whistle, so it is best to put down one or two more answers than the number of marks, unless the number of answers allowed is specified.

Student answer

2. (a)

Make sure you are with at least one other student. Have a mobile phone with pre-programmed numbers to call for help if needed. Have a whistle to blow to attract attention. [2 marks]

Examiner feedback

Two good points are made but the mobile phone and whistle serve the same purpose, so are worth only one mark.

Sample question

3. (a) What should be included on all recording sheets? **[3 marks]**

(b) What is the difference between primary and secondary data? **[2 marks]**

(c) What advantages does objective (quantitative) data have over subjective (qualitative) data? **[2 marks]**

(d) How can secondary data help in a weather study? **[1 mark]**

(e) Give **two** advantages of carrying out a pilot survey. **[2 marks]**

See www.oxfordsecondary.com/esg-for-caie-igcse for the mark scheme for this question.

Analysis

✓ There are many other considerations when planning certain investigations. More will be included in Chapter 32.

Student answer

3. (a)

Spaces for recordings, name of observer and date and time. [2 marks]

Examiner feedback

Very occasionally, when there are a lot of short possible answers that only need to be listed (as here), two points may be required for one mark.

Key ideas

- Data collected should be appropriate to the study.
- It should be collected and recorded accurately.
- Enough data should be collected to form a representative sample and to be statistically acceptable.

Common error

Remember to read the angle off a non-digital clinometer at eye level and that the reading should be between the same heights on the upright ranging poles.

Measuring accurately

Readings from digital instruments are more likely to be reliable than student readings of non-digital instruments, but measuring errors can be reduced by:

- knowing how to read each instrument correctly
- taking an average of three or more readings or measurements.

Decide what equipment will be required, particularly for measuring:

- the channel width, depth, speed of flow and shape and size of bedload
- a beach profile, shape and size of beach material and movement of it along the beach
- weather study
- human studies such as pedestrian or traffic counts, sphere of influence, pollution, urban land use and environmental quality.

Measuring the cross profile of a stream channel

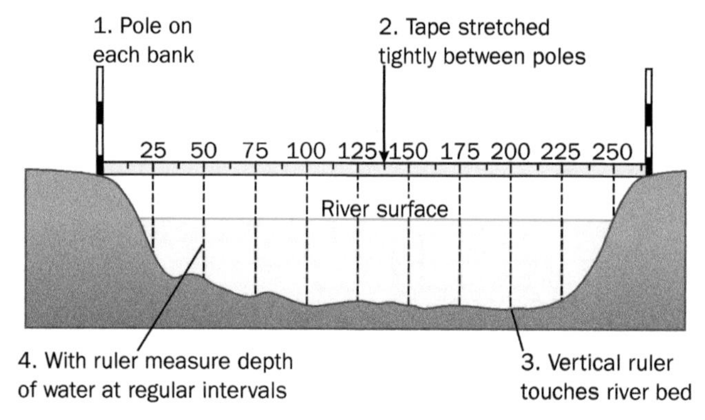

▲ **Fig. 32.1** *The smaller the intervals between measurements, the more accurate the profile will be*

Surveying a slope profile

Use two **ranging poles**, a **clinometer** (preferably a digital one), a prepared recording sheet, a pencil and a clipboard. The ranging poles must be kept vertical and rest on the surface, not sunk in to the ground or riverbed.

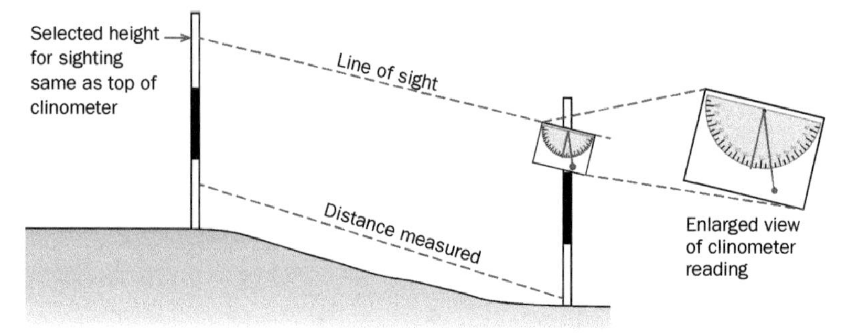

▲ **Fig. 32.2** *This diagram shows measuring the slope at regular distance intervals. It is not as accurate as placing the poles at known breaks of slope and measuring the distance and angle between them. The right hand pole should have been placed a short distance to its left at the break of slope.*

Investigating pebble load size and shape

Measuring the long axis of a pebble with a **calliper** is more accurate than using a ruler and judging it by eye.

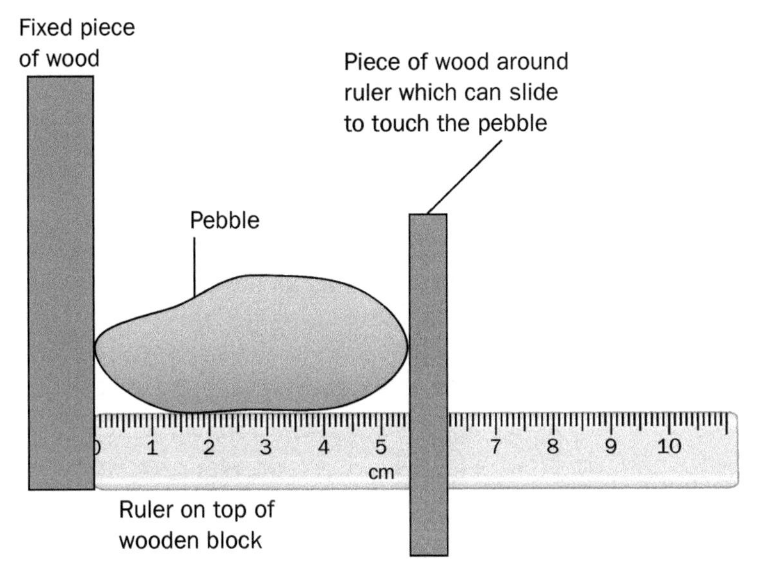

▲ **Fig. 32.3** *A pebbleometer, which can also be used to measure the long axis of pebbles*

The average particle size at each sample point can be determined. The angularity or roundness of pebbles is assessed subjectively by comparing them with drawings in Powers' **Roundness Index Chart**. If a quadrat is used, one pebble could be collected from a set number of squares using either a random or systematic sampling method.

Counting methods

Pedestrian and traffic flows are counted and recorded on a **tally chart** or by clicking an automatic counter as each individual person or vehicle passes.

TRAFFIC RECORDING SHEET

Day: Tuesday Date: November 8th Time: 8–8.10

Street: West Street Site: after junction with Hope Avenue

Inbound/~~outbound~~ side Weather: wet and windy

Mode	lorries	vans	buses	cars	motorcycles	bicycles
Tally	𝍸𝍸 |	𝍸𝍸	||||	𝍸𝍸𝍸𝍸𝍸𝍸 |||	𝍸𝍸𝍸𝍸𝍸 |||	𝍸𝍸 |||
Totals	11	10	4	33	28	13

▲ **Fig. 32.4** *A tally chart is quick to count as recordings are in groups of five*

Each count should be long enough to give a representative sample for reliable data to be collected.

Common error

Remember that counts in different locations must be done at the same time for a fair test.

Questionnaires

It is often important to do a stratified sample for questionnaire enquiries; the number in each age group asked should be representative of the total population. (Details can be obtained from census data.) For many surveys it would be preferable to interview an equal number of males and females to obtain a balanced view.

Afterwards, collate the data collected into data tables.

Land use surveys

In urban areas, systematic sampling can be done as **transects** along roads starting at the town centre and ending at the edge of the CBD or town. The recording sheet may have spaces for land uses on more than one floor of a building. Land use categories are decided beforehand. Observed land uses are plotted on a large-scale plan of the area on which separate buildings are shown.

For rural land use on public land, it is also easy to use point or area sampling, as there are usually fewer obstructions.

Vegetation

Investigate changes in vegetation species and ground coverage by area sampling using a **quadrat** at regular intervals along transects

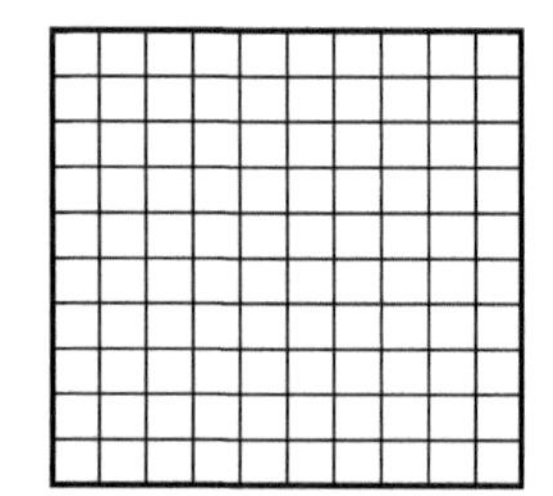

▲ **Fig. 32.5** *Quadrat with 100 small squares*

Bipolar surveys

These use a range of scores to assess environmental quality, for example from zero to four, with four representing the highest quality and zero the worst.

Other scales use negative and positive figures either side of zero; the average situation is represented by zero. Negative figures show the degree of undesirable quality and positive figures indicate the degree of the good aspects of the site.

Scores for several aspects of environmental quality are obtained using a pre-prepared **environmental quality reference sheet** and then totalled to give an overall assessment of each site.

	0	1	2	3	4	
Very low quality of house exteriors						Excellent quality of house exteriors
Very low quality of roads and pavements						Very high quality of roads and pavements
A lot of litter						No litter

Common error

Remember that observations in different locations must be done at the same time for a fair test.

Exam tip

Be able to describe any method of collecting data. You will find this easier if you have practised it in the field with the necessary equipment.

Recording a view in the field

A landscape or scene is recorded by a **field sketch** or by taking a photograph. An example of a field sketch is given in Fig. 30.1 on page 168.

Sample question

1. Describe how you would measure any slope (such as the long profile of the stream bed or the valley side). **[6 marks]**

Analysis

✓ Try to practise the techniques so that you remember better how to do them and will be able to explain the steps in a logical order.

✓ Give detailed explanations of research methods.

Student answer

1. Place a ranging pole at the lower end of the slope and another at the first clear break in slope. Measure the distance between them with a tape measure. Use a clinometer to measure the angle between the height of the clinometer on the lower pole and the same height on the other pole. Move the first ranging pole to the next break of slope and repeat the procedure. [5 marks]

See www.oxfordsecondary.com/esg-for-caie-igcse for the mark scheme for this question.

Examiner feedback

The instruments used to obtain each measurement are clear. It should also be mentioned that the poles must be held upright and that they should be inserted into the ground to the same depth as each other.

Sample question

2. You are to investigate changes in vegetation cover down a slope. What equipment would you use to investigate the vegetation cover? Justify your choice. **[2 marks]**

See www.oxfordsecondary.com/esg-for-caie-igcse for the mark scheme for this question.

Analysis

✓ The method for measuring the slope can be used on many landforms.

✓ In addition, you could be asked how you would design a suitable recording sheet and how many people you would need to carry out the investigation.

✓ Recording the result is omitted as it is not part of the *method* of the research.

Sample question

3. Look at the random number table.

39	26	02	11	98	55
58	07	46	60	77	04
17	83	29	32	41	36
48	65	08	93	55	69

(a) Use the random number table, reading vertically down, to identify the first **five** pebbles you would pick up when investigating pebbles along a line on a beach. **[1 mark]**

(b) Compare **one** advantage and **one** disadvantage of taking photographs or drawing field sketches to record features of a landscape. **[2 marks]**

See www.oxfordsecondary.com/esg-for-caie-igcse for the mark scheme for this question.

Analysis

✓ Any randomly organized list of numbers could be used instead.

✓ On the Alternative to Coursework examination paper one of the two questions is based on one of the topics of physical geography in the syllabus. The other question is based on human geography. Rivers, coasts, weather and vegetation will be more likely to be the subjects of the physical-based questions than earthquakes and volcanoes.

Sample question

4. (a) How can total environmental scores for each survey site be used? **[2 marks]**

(b) Why should environmental quality surveys be done at the same time at each site? **[1 mark]**

See www.oxfordsecondary.com/esg-for-caie-igcse for the mark scheme for this question.

Analysis

✓ If the research involves a person giving a subjective score, an average of the scores of at least three people will eliminate or reduce subjectivity.

✓ Ranking in order is a useful way of showing comparability.

Raise your grade

Sample question

5. (a) Apart from planning the questionnaire sheet, what other planning should be done for a class questionnaire enquiry about the views of the population as a whole? **[4 marks]**

(b) Why should the times and day be chosen carefully when conducting investigations such as questionnaires and most traffic or pedestrian counts? **[2 marks]**

(c) Why is the question 'Do you live locally?' a common first question on a questionnaire? **[2 marks]**

(d) Why is it more appropriate to ask people which age range or which income range they are in, than ask for the exact figures? **[1 mark]**

(e) What advice is a teacher likely to give students about how to conduct a face-to-face questionnaire to avoid and deal with problems? **[2 marks]**

(f) A student suggested that a questionnaire about tourism on an island should be asked of people waiting in the queues to check in at the airport for their flights home. State **one** advantage and **one** disadvantage of her proposal. **[2 marks]**

(g) A teacher advised students doing a survey about the general population's views to interview five people, one from each predetermined age group. The teacher told them to ask the fifth person that passed after each interview had finished until they had managed to ask one person in each of the age groups. What kind of sampling did the teacher recommend and why was it a suitable method? **[4 marks]**

Analysis

✓ In a question asking for advantages and disadvantages, make it clear in your answer which is the advantage and which is the disadvantage.

✓ In order to help you answer questions about unfamiliar things, imagine yourself at the scene or in the situation doing the task.

Student answer

5.a. Decide on the type of sampling to be used, and how to stratify the sample to ensure that an equal number of men and women are sampled. Decide whether to do a doorstep survey or interview people in busy parts of town, such as the CBD. [3 marks]

Examiner feedback

There are 4 marks allocated for the question but the answer only has three points, so the maximum mark that could be gained is 3 marks. The obvious omission is how many to sample (30 or more).

Sample question

6. (a) Describe methods you would use to test a river for pollution and to observe
indicators of water quality. **[5 marks]**

(b) Explain where you would take samples in order to find out the source of pollution. **[2 marks]**

(c) Describe how you would find out the spheres of influence of **two** supermarkets. **[2 marks]**

See **www.oxfordsecondary.com/esg-for-caie-igcse** for the mark scheme for this question.

Analysis

✓ In each of the two questions set on the Alternative to Coursework examination paper, 12 marks out of the total of 30 are allocated to techniques for observing and collecting data.

Student answer

6.a. Get a sample of water from the river to take back to the school laboratory for testing. Observe the colour of the water. See if there are any live or dead fish in the river. Put your hand in the water to see how deep it is when it can't be seen. [2 marks]

Examiner feedback

This is a poor answer. More detail is needed about how to take a sample of water, such as what it is collected and transported in.

A mark would be given for noting the colour of the water and a second mark for observing signs of fish and animal life or death.

The suggestion that a researcher testing for suspected pollution should put their hand in the water without first putting waterproof gloves on is not sensible and would not obey health and safety guidelines.

Key ideas

- Data should be presented in an appropriate form.
- All appropriate labels, a key and a title should be given.
- Graphs should be titled and their axes titled and labelled.
- Data should be analysed to identify trends, patterns and other findings of significance.
- The significance of data to the investigation should be indicated and explained.
- A conclusion to the investigation should be written.
- The key pieces of evidence to support the conclusion should be included.

Presentation, analysis and interpretation of the data

Simple statistical analysis

- **Rank order** indicates the relative importance of the data.
- The range of the data shows its spread or distribution.
- Mean, **median** and **mode** indicate the middle of a data set.
  - The mean is a good indicator if there is a **normal distribution**, and can be used in further calculations.
  - The median should be used if the distribution has extreme values at one end of the range.
  - The mode has limited use, but the modal class of a histogram can be a valuable indicator.

Example data set

Weekly total sunshine hours at a place over a nine-week period

Hours (arranged in rank order):

64

60

53

44

35 → median

33

24

24

21

total = 358
mode = 24
range = 21 to 64 = range of 43
mean = 358 divided by 9 = 39.8

Describing trends

- Comment on increases and decreases, both over the whole time period and the pattern within it.
- Describe in detail, such as *slight* decline, *large* increase and support your description by quoting the calculated amount of increase or decrease in figures or words (it 'halved' or 'tripled', etc.) and the period of time over which it occurred.
- Always refer to both axes of a graph.

Exam tip

Know the advantages and disadvantages of constructing and using data tables and different kinds of maps, diagrams and graphs (Chapters 28–30).

Common error

Remember that simply listing figures, for example the values for each year in turn, is not an acceptable method of describing trends. The figures must be interpreted and used.

Patterns, relationships and anomalies

An example of a pattern would be noticing from data that no rain falls when cloud cover is 4 oktas or less and most rain falls when the cloud cover is 8 oktas, suggesting a positive relationship between the two variables.

- Patterns from data in ranked tabular and graphical form can be used to deduce relationships.
- A description of patterns on a map involves using words such as 'mainly', 'least' and 'more', while expressing the distribution of the feature in question. It is useful to indicate locations using compass points and features such as hills and valleys.
- It may also be appropriate to refer to shapes, using terms such as 'round', 'elongated', 'linear', or spread using 'nucleated', 'dispersed', etc.

Scatter graphs (Chapter 30) are valuable for recognizing whether or not a relationship exists between two variables and whether it is a **positive** or **negative (inverse) relationship**.

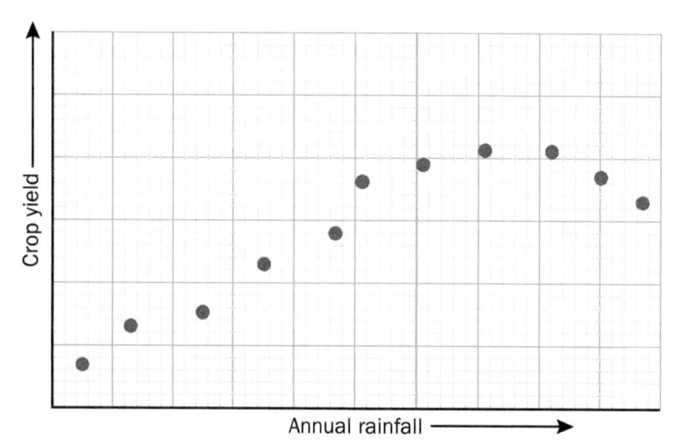

▲ **Fig. 33.1** *In some data sets both trends can be found, as in this graph. (The values on the axes have been omitted because they vary according to the rainfall requirement of the crop.)*

Suggest explanations for the results and make conclusions

Use your knowledge to explain the distributions, trends, patterns, relationships or anomalies or suggest why none were found.

Refer to the aims of the investigation and the hypothesis being tested in your conclusion. It is important to look back at this when answering questions or making conclusions.

State whether or not the evidence allows you to accept, reject or partially accept the hypothesis. Quote data to support your decision.

To what extent was it reliable?

A good sample will be as representative of the population as possible and will avoid bias. It is acceptable to describe it as a 'fair test' or 'reliable', provided that you explain why it is so. Do not describe it as 'accurate'.

Several anomalies suggest that the hypothesis is partly but not wholly true. Support this conclusion by referring to the anomalies.

Key ideas

- The investigation should end with an evaluation stating which methods were successful and which less successful or unsuccessful.
- Ways in which the investigation could be improved should be identified.
- Suggestions for further enquiries to extend the work should be suggested.

Evaluating the investigation

To what extent was it successful?

- Was the hypothesis valid?
- If the hypothesis was sensible and the investigation well planned, it should have been generally successful. If it is only partly true, was enough data collected for the hypothesis to be tested? Was any data missing?

Was the data collected reliable and how could the investigation be improved?

- Was the investigation affected by bias?
- Did the time or day chosen for the survey cause bias by preventing certain groups of people from being included?
- Were the investigating techniques the most appropriate? Would a different sampling method have been likely to give better results?
- Was the sample size too small (less than 30)?
- Were the results affected by an unexpected factor?
- Were questionnaire results possibly affected by people not answering honestly because it might make them look bad?
- Could student error in measuring or reading an instrument have caused an anomalous result?

After identifying a possible fault, suggest specific ways in which it could be corrected. For example, a study could often be improved by repeating it using better equipment, such as digital measuring equipment, which would eliminate possible reading errors.

Common error

Remember to give detail and avoid making vague statements, such as suggesting that an improvement could be to 'do more studies'. This is without value unless you state the type and/or location of the extra studies.

How could the investigation be extended?

Suggestions must be practical.

- Comparison with a past study or with a study obtained from a secondary source would be valid.
- Suggest other hypotheses about the subject which could be used to widen the research.
- Suggesting doing more of the same is not acceptable because it does not suggest an extension.

Exam tip

Be able to analyse data sets and support conclusions you make about the hypothesis being tested and its reliability.

Sample question

1. Students at a school on the west coast, with prevailing onshore SW winds from the sea, did the following investigations.

(a) While investigating the hypothesis: ***Diurnal temperature ranges are greatest when atmospheric pressure is highest***, students presented on a scatter graph the temperature and pressure data they had collected.

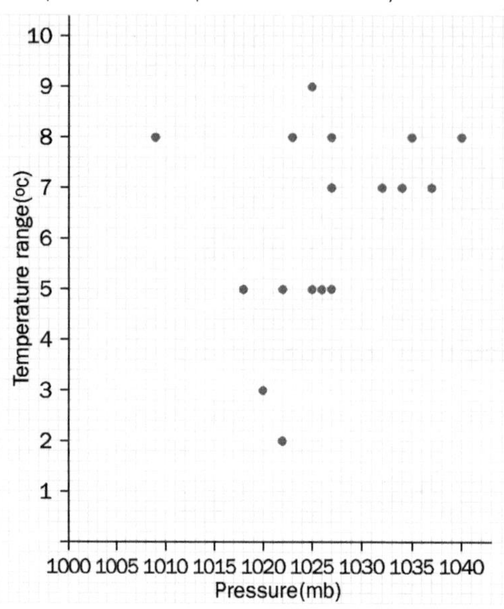

What would the students conclude from the scatter graph about the hypothesis? Use data to support your view. **[2 marks]**

(b) To investigate their hypothesis: ***Precipitation is greatest when winds are from a westerly direction*** the students drew dispersion diagrams for each wind direction.

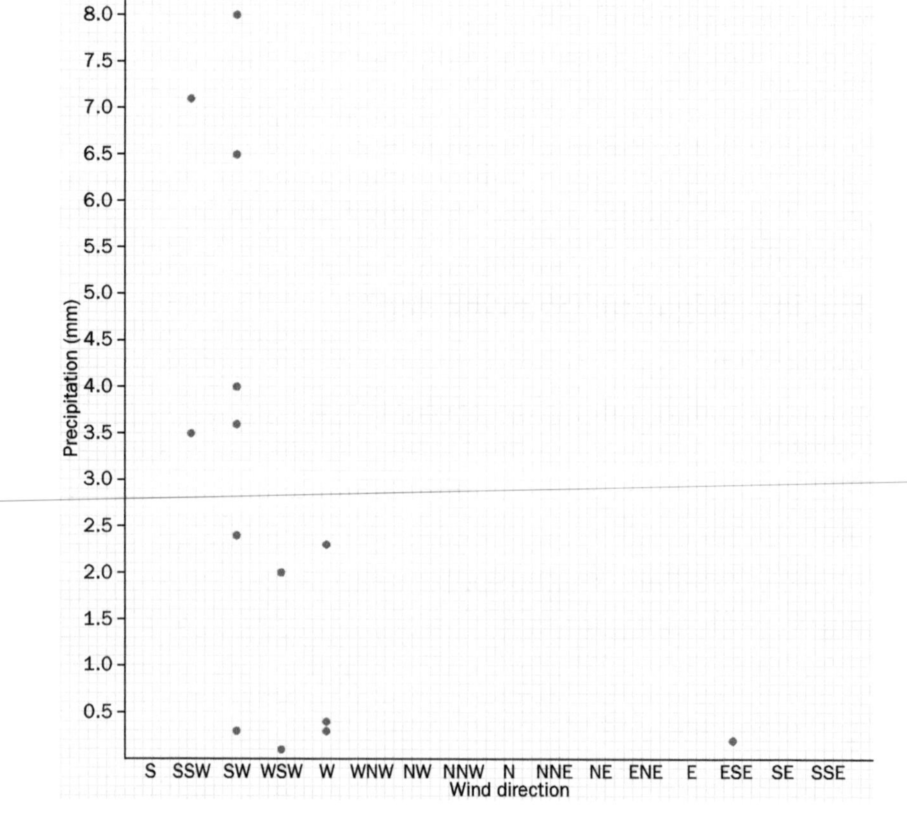

What conclusion would the students make about whether or not the hypothesis is true or false? Support your conclusion. **[3 marks]**

(c) Which result is an anomaly? How can it be explained? **[2 marks]**

(d) Explain the results you obtained in part (b). **[2 marks]**

(e) Describe **two** ways in which the investigation could be extended. **[2 marks]**

Analysis

✓ See Chapter 30 for more information on data presentation.

✓ The phrase 'use data' means that you should select it from the resource to illustrate your comment.

✓ When referring to data in graph form, remember to quote the values from both axes.

✓ The phrase 'explain your results' means that you should explain why you would have expected the conclusion you reached.

✓ When stating how an investigation could be extended, do not suggest more of the same. Suggest a related investigation that would complement the study done, not the same one with more of the same type of data.

✓ An anomaly does not fit the relationship. It is most easily recognized on a scatter graph as a point considerably further from the best-fit line than the others.

See www.oxfordsecondary.com/esg-for-caie-igcse for the mark scheme for this question.

Student answer

1.b. Only one day with rain comes from the ESE. On all the others, winds come from the SSW, SW, WSW and W, so the hypothesis is true. [2 marks]

Examiner feedback

One more supporting reason is needed as the mark allocation is 3. It is preferable to try to quote supporting statistics when answering such questions, such as, the day with an ESE wind has only 0.2 mm of rain, a very small amount, or that it is only one of 13 days with rain.

Sample question

2. Students were testing the hypothesis: ***The amount of benefit tourists bring to the economy varies with their ages***.

(a) The students conducted a questionnaire with 100 tourists. Question 6 of their questionnaire was: 'What type of accommodation are you staying in?'
The students presented the answers to this question in a data table.

Answers to question 6					
Age group	Apartment	Caravan/ camping	Hotel	With friends/ family	Total
Under 20	9	10	0	1	20
20–34	12	7	1	0	20
35–49	12	2	6	0	20
50–64	9	0	10	1	20
65 and over	5	0	13	2	20
Total	47	19	30	4	100

Rank the types of accommodation from most to least popular. **[1 mark]**

(b) What general conclusions can be drawn about how accommodation varied with age groups? **[4 marks]**

(c) State **one** other investigation the students could have done to test the hypothesis. **[1 mark]**

(d) Describe how you would present this data in diagram form. **[4 marks]**

(e) The students also asked the same 100 tourists the question, 'How did you travel during the longest part of your journey to get here?'. They presented their results in a bar graph.

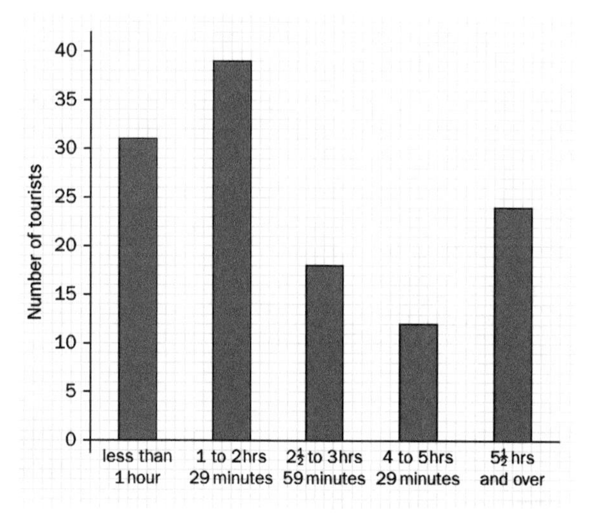

Hypothesis 2 was: ***The number of tourist visitors decreases with increasing travel time to the resort***. Do you agree with this hypothesis? Quote data to support your conclusion. **[3 marks]**

Analysis

- ✓ For parts (b) and (e), support statements by quoting data. It is essential to give reasons for conclusions.
- ✓ One advantage of interviewing 100 people is that the results will be percentages, which saves calculations for analysis of the data.

See www.oxfordsecondary.com/esg-for-caie-igcse for the mark scheme for this question.

Student answer

2.b. Camping was used by 10 people aged below 20 out of the total of 19 people who camped. Most of the tourists interviewed stayed in apartments. Hotels were also popular, as 23 out of 30 using hotels were over 49 years of age and only 7 were younger.

[1 mark]

Examiner feedback

The statement about camping is weak and restricted to one age group when the question asks about *variations* with age *groups* and asks for *general* conclusions which means you need to look at the overall picture. The most obvious general conclusion to make is that there is a steady decline in the popularity of camping with each older age group.

The second sentence would score no marks because it does not mention age at all. Mention of the popularity of hotels is irrelevant to the question because it is not linked to age groups, but the rest of the sentence does make the point in the mark scheme that hotel use increased with age.

Student answer

2.d. Draw five bar graphs side by side for each age group, one bar for each accommodation type. Shade each accommodation differently and add a key. [2 marks]

Examiner feedback

This is a poor method because it would show the information in 25 bar graphs. The purpose of a diagram is to give a quick visual impression of a situation. It would take much longer to see patterns which would be more immediately obvious on five divided bar graphs or pie graphs.

One mark would be awarded for a method that would work and another for the information about the key.

Student answer

2.e. The hypothesis is false. The longest journey time for most tourists was 1–2 hrs 29 minutes, whereas only 12 travelled for 4–5 hours 29 minutes.

[1 mark]

Examiner feedback

The conclusion is correct but the reasoning is poor. Be careful with the use of the words 'most' and 'fewest'. It is not true that *most* tourists had that travel time as the number of tourists with that journey numbered only 39 out of the 100 asked. For 'most' to be correct the number would have to 51 or over.

Also the figures quoted are simply *repeating* information given in the graph not *using* them to prove the point that there is no clear relationship.

Sample question

3. An area has four plant species: P, Q, R and S. Four line transects were taken to investigate the vegetation in the area.

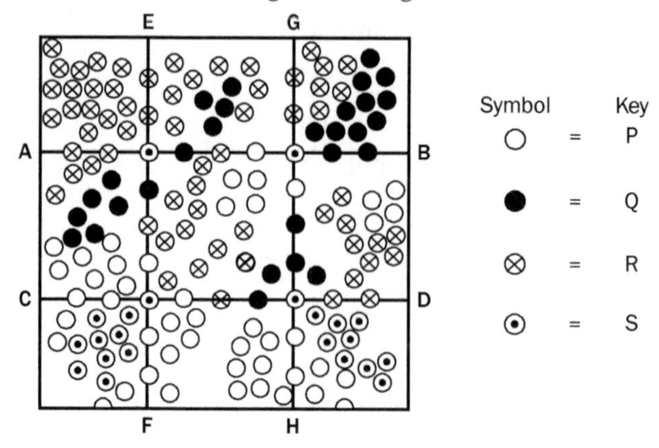

(a) List the plant species on the transect line from G to H. **[1 mark]**

(b) What do the results from this transect survey suggest about the most abundant vegetation species in the area? **[1 mark]**

(c) Use the diagram and your answer to part (b) to explain why line sampling gives unreliable results in some circumstances. **[3 marks]**

(d) Why should you not describe the results of any sample as 'accurate'? **[1 mark]**

See www.oxfordsecondary.com/esg-for-caie-igcse for the mark scheme for this question.

Analysis

✓ The phrase 'use the diagram' means that you should quote evidence from it where appropriate.

✓ A successful sample can be described as a fair test or a reliable test, but never as accurate.

Sample question

4. (a) What type of sampling is shown in the diagrams? **[1 mark]**

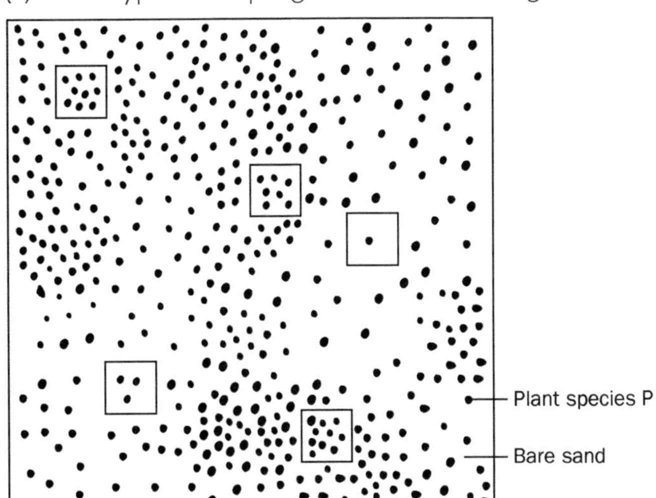

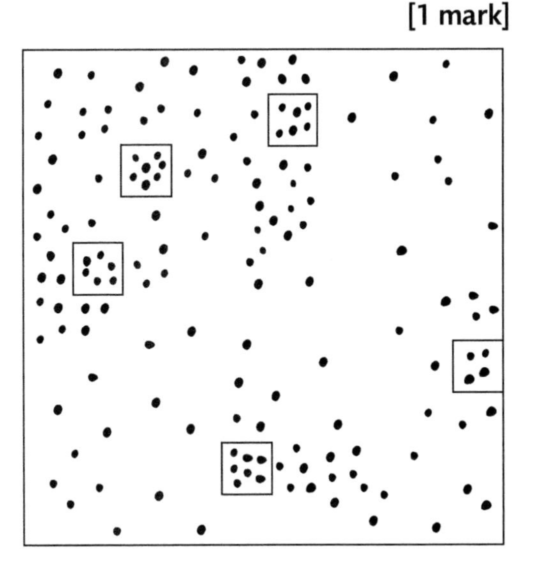

(b) Each sample is 1 metre square and the whole area is 10 metres by 10 metres. Estimate the total number of plants of species P in the whole area on the left-hand diagram.. Show your working. **[3 marks]**

(c) The right-hand diagram shows the same sand dune area after it has been opened up to tourists. The estimate calculated from the samples shown is the same as the result in (b) for the total of plant species P for the whole area. Compare the results of these two samples with what you see in the two diagrams. **[1 mark]**

(d) Suggest why this sampling technique failed to show the effects of the tourism and suggest a type of sampling that would have worked well to show the difference. **[2 marks]**

(e) List ways in which results of samples and surveys can be improved. **[2 marks]**

Analysis

- ✓ These questions illustrate the importance of choosing the most appropriate method of sampling for the situation when investigating distributions over an area.
- ✓ Random sampling is a bad choice if some of the features are in groups, but it is safe to use if there is a fairly even spread over the area.
- ✓ You may be asked to criticize sampling choice and should support your answer with evidence.
- ✓ If you count the number when asked to estimate, not only would you waste valuable exam time, but your answer would be incorrect because the task did not ask you to calculate.

See www.oxfordsecondary.com/esg-for-caie-igcse for the mark scheme for this question.

Glossary

AIDS Acquired Immune Deficiency Syndrome (AIDS) is an infectious disease caused by the human immunodeficiency virus. It causes an inability to fight infections that frequently cause death

Accessibility How easy it is for people to travel to and from a particular point, usually in terms of time travelled

Acid rain Rain containing sulfuric and nitric acids from gases released from power stations, industry and vehicle exhausts

Age–sex pyramid A diagram which shows males and females in a population using horizontal bars, which can show numbers or percentages. Each bar represents a five-year age group

Agriculture Farming; the artificial cultivation of plants (crops) and rearing of animals for food and other products

Altitude The height above sea level

Altocumulus Middle-level cloud made of water droplets; can be white or light grey if thick

Altostratus Middle-level cloud made of water droplets; can be thin and white, or thick and grey

Aquifer A layer of rock which contains water

Arable farming Growing crops

Area of origin of migrants Where the migrants come from, causing its population to decrease

Ash Fine dust produced by a volcano; the finest pyroclastic material

Aspect When used in geography, this means the compass direction that a slope faces

Assembly The putting together of component parts to make a product

Atmosphere A mixture of gases that encircle the Earth: mainly nitrogen and oxygen with some others, e.g. water vapour, methane and ozone

Attrition Process of erosion in which fragments of the river's/wave's load become smaller and more rounded by collision with the river bed/banks/coastal rocks/each other

Beach Sand and shingle between low- and high-water marks. Shingle forms a steep slope, whereas sand beaches are gently sloping. Beaches can be straight, crescent-shaped in bays, and triangular at the head of inlets

Biodiesels Made from vegetable oils, such as rapeseed oil, but also from recycled cooking oils from restaurants and kitchens

Bioethanol An alcohol made by fermenting the sugar in plants, such as maize. Bioethanol can be used as a fuel for vehicles

Biofuels These are fuels from biomass. They include liquid fuels (bioethanol and biodiesel), biogas and solid biofuels (including fuelwood)

Biogas Methane produced by the breakdown of organic material by bacteria. It can be produced either from waste materials or by the use of energy crops

Birth rate The number born each year per 1000 people

Borehole Like a well, but just a few centimetres wide. Drilled into the rock so that groundwater can be pumped out

Braided Continuously splits into smaller channels (distributaries) and re-joins

Calliper An instrument with two arms ending in points which are adjusted to fit across the stretch to be measured. The calliper is then placed against a rule to read off the distance

Carbon footprint The total amount of greenhouse gases, including carbon dioxide, produced by a country or individual

Carbon sink An absorber and store of carbon

Cash crop A crop sold for money

Cirrocumulus Slightly heaped, high cloud made of ice crystals

Cirrostratus High cloud made of ice crystals; it has a wide horizontal extent and often covers the whole sky

Cirrus High cloud made of ice crystals; narrow wisps or threads; it can also be feather-like

Clinometer A device that measures angles

Commercial farming The farmer sells his or her output to make a profit. This is typical of modern, large-scale farming

Community-based tourism Tourism organized after consultation with the local community to benefit them with employment as tour guides and arranged paid accommodation for tourists in their homes

Commuting Travelling to and from work, usually from home on a daily basis. Commuting usually implies that the journey is not extremely short

Comparison shops Sell goods not bought every day. The shopper visits more than one store to look at different prices and quality, for items such as furniture or shoes

Confluence The point where two rivers meet

Conservative plate margin A plate margin where plates slide past each other and a plate is neither created nor destroyed

Constructive (divergent) plate margin A plate margin where plates diverge and a new oceanic plate is created

Constructive waves Gentle waves that build up the beach because their swash is stronger than their backwash

Contours Lines on a map which join points of equal height above sea level

Conurbations Cities which have grown outwards and merged with other towns and cities to produce very large settlements

Convectional rainfall Rain that falls as a result of the cooling of air made to rise by heating

Convenience shops Sell goods bought almost every day from local shops, such as bread and milk

Coral polyp A tiny marine animal with a calcareous skeleton

Coral reef A hard, rocky, calcareous colony of millions of coral polyps joined together. New generations of polyps grow on top of old ones, so the reef grows upwards and outwards as the corals compete for food

Corrasion (abrasion) Process of erosion in which the bed and banks of the river are eroded by the abrasive effect of the river's load; or by sediment eroding solid rock as it is thrown against it by powerful waves

Corrosion (solution) Process of erosion in which certain minerals in rocks are dissolved by the river/sea water

Counter-urbanization The movement of population from towns back to rural areas

Crater The depression at the top of a volcano

Cumecs Cubic metres per second, used to measure discharge

Cumulonimbus Low-level cloud made of water droplets. From a low base the cloud extends up to high levels. It is a dense, dark grey cloud with a great vertical extent. It grows from a cumulus cloud to have a high, billowy head (or a flat top if it reaches the tropopause). If it then spreads out, it has an anvil top. It is composed of ice crystals at the top and water droplets at lower levels. It produces very heavy rain, or snow showers, often with hail, thunder and lightning

Cumulus Low-level cloud made of water droplets. White with a darker, flat base and globular upper surface. It may have a small or considerable vertical extent

Death rate The number who die each year per 1000 people

Demographic factor An influence of the size or structure of the population

Dependency ratio The total of young and elderly dependents divided by the number of working age and multiplied by 100. The calculation gives the number of people who have to be supported by 100 workers

Dependent population People not of working age, often taken as 0–14 and 65+ years, and are supported by the workers

Desalination The removal of salts from saline (salty) water to produce water suitable for human consumption. This is generally seawater, but some groundwater can be saline

Desertification A series of processes that change once productive land into land which can no longer support vegetation and crops so looks like a desert

Destination area of migrants Where the migrants move to, causing its population to increase

Destructive (convergent) plate margin Where two plates converge and one is destroyed

Destructive wave Powerful waves that attack and weaken rocky coasts and sweep loose material away by breaking with a steep plunge. Their backwash (water moving back to sea) is steeper than their swash (water that moves up the beach)

Development The way that a country becomes more advanced in its economy, infrastructure and the economic and social wellbeing of its citizens

Discharge The volume of water flowing down the river at any time

Domestic tourist A person taking a holiday in her/his own country

Dominant wind The direction of the strongest winds

Dormitory village A small settlement where many of the people commute to work in another settlement

Drainage Rivers, streams, lakes and their features. They are usually shown in blue on a map

Drainage basin The area drained by a river and its tributaries

Drainage density The total length of the rivers and streams in an area in kilometres divided by the area in square kilometres

Drought A longer than usual period of dry weather, which occurs at a time when more rain is expected to fall

Dry point Higher points in otherwise poorly drained areas

Eastings Lines running north to south on a survey map

Economic factor An influence connected with the production or use of wealth

Economic migrants People who choose to move for better jobs and higher pay

Economically active population The population of working age that contributes to the wealth of the family and country

Ecosystem An area in which plants and animals live in balance with their environment (climate, soil and vegetation) and are interlinked with it

Ecotourism Tourism organized to allow tourists to visit while preserving the environment

Emigrant A person who leaves the country or area

Employment structure The proportion of people working in primary, secondary, tertiary and quaternary activities in any country or region

Enhanced greenhouse effect The increase in greenhouse gases in the atmosphere as a result of their release during human activities. It is believed to be the reason for the recent increased rate of global warming

Environmental quality reference sheet A list of agreed criteria for each score to be used during a bipolar survey by each research group, so that scoring is as objective as possible

Environmental refugees People who flee from environmental disasters, such as drought

Epicentre The point on the Earth's surface directly above an earthquake focus

Extensive farming Inputs and outputs per hectare are low and overcome by using a large area of land

Falling limb The part of the hydrograph where discharge is decreasing

Fault A crack in the rocks of the Earth's crust where the rocks move and are displaced

Field Used in research to mean where it is undertaken (it may be indoors as well as outdoors)

Field sketch A quick drawing of the main features of a landscape that can be annotated

Finite resource A resource that cannot be renewed when it has been used up

Floodplains The area of flat land either side of a river, affected by flooding

Focus The point in the Earth where an earthquake occurs

Food web When individual food chains link together because an animal has more than one predator

Fossil fuel Organic material (plants and animals) which was growing millions of years ago. To grow, this organic matter got its energy from the sun. When we use these fuels, we are using the sun's energy from millions of years ago which has been stored in the fossil fuels

Gabion Wire cages filled with stones that can be wired to others and used as a groyne or seawall

Geothermal power Using the heat contained deep in the Earth which is more concentrated in volcanic areas

Ghetto Part of an urban area where there is a concentration of a single racial group living in relative poverty

Global footprint he quantity of carbon dioxide that is added to the atmosphere by a person's or country's activities

Global warming The increase in the temperature of the atmosphere; it has increased rapidly since 1980

Globalization This is the growth of international integration, in other words, the increase in links between different parts of the world and different countries

Greenhouse gas An atmospheric gas that allows the passage of incoming solar radiation through to reach the Earth's surface, but absorbs outgoing radiation from the Earth

Grid reference A system which allows locations on a map to be described precisely by a four- or six-figure reference

Gross Domestic Product (GDP) per capita The total value of the goods and services produced by a country in any one year and divided by the population of the country. It is expressed as $US per person so that countries can be compared

Groundwater Water held within the spaces in porous, permeable rocks

Groundwater flow Water flowing through the rock

Groyne A barrier, usually made of wood, placed on a beach at right angles to the shore to catch sand and shingle as it is moved by longshore drift

Gullying Removal of the soil by water concentrated in channels, known as gullies

Hazard A threat that could injure people or damage property and infrastructure

High-technology industry Production that uses the most advanced technology to make the products

Hinterland An area linked to the port from which goods are exported and to which goods are imported

HIV Human Immunodeficiency Virus that attacks the body's immune system causing it to be unable to fight infections

Housing Size, number of storeys, building materials, quality, windows, the building plots

Human Development Index (HDI) A composite index. It takes into account a country's: GDP per capita, adult literacy, educational provision and life expectancy at birth

Human features Features of buildings and settlements, agriculture, industry, transport

Humidity The amount of water vapour in a given volume of air

Hydraulic action (coastal) Air compressed in cracks by waves expands violently as the wave retreats, enlarging the crack

Hydraulic action (river) Process of erosion by which the force of running water alone removes material from the bed and banks

Hypothesis A statement that can be proved or disproved by being tested

Illegal migrant A person who moves into a country without permission

Immigrant A person who comes into the country or area

Independent population People of working age, often taken as 15–64 years

Industrial zone An area with a large amount of industry

Infiltration Water entering the ground surface

Input Something that is needed, e.g. by industry or farming to function

Insolation Incoming solar radiation

Intensive farming Inputs and outputs per hectare are high

Interception Rain which is caught by vegetation before reaching the ground

Inter-Tropical Convergence Zone (ITCZ) A broad area where winds meet between the tropics

Isobar A line joining places with the same pressure

Isohyet A line joining places with the same rainfall

Isoline A line joining places with the same value

Isotherm A line joining places with the same temperature

Key A list of the symbols shown on a map which shows what they mean Scale How distance on the ground has been represented on a map

Lag time The difference in time between the highest rainfall and the highest discharge

Lagoon An area of shallow water separated from the sea, e.g. by a coral reef

Lahar A powerful current of mudflow – water mixed with mud, rocks and boulders (material from volcanic eruptions) – running down a river channel, valley or slope

Large-scale map A map which shows a relatively small area of land in great detail, for example a topographic survey map

Lateral blast When a volcano erupts sideways with great force, producing gas and pyroclastic material

Lava Molten rock on the Earth's surface

Leisure tourist A tourist away from home for a holiday

Load The boulders, pebbles, sand, silt, mud and minerals in solution carried by a river

Long-haul Travel to a destination in more than three hours

Magma Molten rock below the Earth's surface

Magma chamber A huge chamber, several kilometres across, which stores molten rock beneath a volcano

Magnitude The total amount of energy released by an earthquake

Mangrove swamp Dense forest of mangrove trees, up to 15 metres tall, growing on mudflats in sheltered tropical waters. They are crossed by many tidal channels swept clear of mud

Manufacturing The production of new products from raw materials by industrial processes

Mass migration The movement of very large numbers of people from one area to another

Mass tourism An influx of very large numbers of tourists to a destination

Meandering Flows in broad bends

Median The middle number in a distribution placed in numerical order

Mega-city A city with a population of over 10 million people

Mercalli Scale A 12-point scale which describes the effects of an earthquake

Migration The movement of people from one place to another

Millionaire city A city with a population of over 1 million people

Mixed farming Growing crops and rearing of animals

Mode The most frequently occurring number in a data set

Monoculture Growing the same crop on the same plot repeatedly

Natural population change The increase or decrease of a population over a given time because of a change *only* in the birth or death rate or both

Negative (inverse) relationship As one variable increases, the other decreases

Net migration The number of immigrants minus the number of emigrants in an area

Nimbostratus Low-level cloud made of water droplets. The base can be low or medium altitude. It is a thick, dark grey layer cloud which produces steady rain

Non-renewable energy Energy we are using faster than it is being created; it will run out

Normal distribution Data with no extreme values on one side which would distort the mean

Northings Lines running east to west on a survey map

Nutrients Minerals that are plant foods; they are taken up by the plant's roots and returned to the soil if the plant decays into it

Organic matter Plant and animal matter

Output What comes out of an industrial or farming process, waste produced during processing and the product(s) for sale

Overcultivation (overcropping) Growing too many crops on a plot so that the nutrients are depleted

Overland flow Water flowing over the ground surface

Over-population There are too many people to be supported to a good standard of living by the resources of the country

Package holiday A holiday for which travel and accommodation are arranged by a tour company or travel agent

Parasitic cone A small cone on the side of a volcano

Pastoral farming Rearing of animals

Peak discharge The highest discharge after rain

Photochemical smog Fog containing nitrogen oxide and other pollutants. A chemical reaction between them takes place in warm sunlight to form ground level ozone

Physical features Relief, drainage, vegetation

Pilot study (practice survey) A small practice investigation using the methods to be used in the full study to be sure they work and are likely to provide the information needed. Changes can then be made if necessary

Plateaux Land which is high and flat

Population density How many people live in a unit area, usually a square kilometre

Population distribution How the population is spread in an area

Population structure The proportions of young, middle-aged and elderly males and females in a population

Porous Describes soil and rock which has spaces through which air and water can pass

Positive relationship As one variable increases or decreases, the other one does as well

Precipitation Liquid and solid particles that fall from the atmosphere to the Earth's surface

Pressure The 'weight' of a column of air in the atmosphere

Prevailing wind The most frequently occurring wind

Primary industry Industry involved with the collection or production of natural resources, food and raw materials directly from the land or sea

Process Activity that occurs, e.g. to make product from raw materials or to produce farm products

Processing The preparation of a raw material into a different state for several purposes

Pull factor Something that attracts people to migrate to it

Push factor Something that people try to escape from by migration

Pyroclastic Refers to the solid material produced during a violent volcanic eruption

Pyroclastic flow A fast-moving current of hot gas and rock, which moves away from a volcano at up to 700 km/h. They normally hug the ground and travel downhill under gravity

Quadrat A square frame, usually a square metre or 0.5 metre square, divided into squares (often 100), so the number of relevant squares can be counted to find the percentages

Quality of life To do with the things that affect a person's well-being and happiness

Quaternary industry Modern, high-tech manufacturing and service industries

Questionnaire A sheet containing a prepared series of questions to use when researching people's opinions

Radiation The transmission of heat by waves from a hot body. The sun emits short-wave radiation and the Earth emits long-wave radiation

Rain shadow The lee slope of high ground and the area nearby with little rainfall

Random sample Data collected in the order obtained from a random number table, so that it is in no particular order

Range The maximum distance that people are prepared to travel to access a particular service

Ranging pole A pole used for surveying which has one pointed end and coloured bands at different known heights

Rank order Data is placed in order of importance from largest or most important (ranked 1) to smallest or least important

Rate of natural population change Birth rate minus death rate expressed per 1000 or as a percentage per year. It can be positive (increase) or negative (decrease)

Raw materials Substances that will be converted into a finished product

Refugees People who have to move to avoid persecution, death or extreme hardship

Relative humidity How much water vapour the air is holding compared with the maximum amount that it could hold at its temperature

Relief The height, steepness and shape of the ground surface

Relief rainfall Rain caused by uplift of air as it rises up the windward side of high ground

Remote sensing Using scientific instruments on satellites above the Earth to provide information about the Earth

Renewable energy Energy which is being produced in nature as fast as we are using it; it will not run out

Representative fraction A way of describing the scale of a map. It shows how many units of distance on the ground are shown by one unit of distance on the map

Reservoir (dam) A barrier on a river to create a store for water to be used during dry periods

Resource Something that is useful to people

Revetment Sloping wall of concrete, wood or boulders placed in front of a cliff

Richter Scale A scale measuring the magnitude of an earthquake

Ridges A long, narrow area of high ground, rather like the spine of an open, upturned book

Rip-rap A pile of large blocks of rock placed in front of cliffs or a seawall

Rising limb The part of the hydrograph where discharge is increasing

Roundness Index Chart A series of drawings of shapes ranging from very angular through angular, subangular, subrounded and rounded to well rounded

Route centre (nodal point) Where roads or other routes meet

Rural Associated with villages and the countryside

Saltation Process of transport in which material moves downstream in a series of hops

Sample A group selected from a larger 'population', where population means the whole of whatever is being sampled for investigation

Sand Small resistant mineral grains, less than 2 mm in diameter, found on beaches

Sand dune A sand hill or ridge of sand parallel to the shore, usually found with other parallel dune ridges. They are deposited by wind

Saturation When the wet- and dry-bulb thermometer readings are the same, the relative humidity is 100% as the air is holding as much water vapour as it can at that temperature

Scarps A scarp is broad, steep slope; it could be the sides of a plateau or a ridge. The slopes may include cliffs

Scrubber A device in a chimney which uses a sorbent to remove harmful substances from the emissions passing through it

Secondary data Information that has been collected by another person not involved in the study or a published source, such as a newspaper, TV report, etc.

Secondary industry Industry involved with processing, manufacturing and assembly of the products we need

Service Anything that is provided in a settlement for the population, including goods that can be bought in shops and other retail outlets

Settlement Features of the buildings themselves (as listed for housing below), the types of buildings, the use of the buildings, the spacing of the buildings and whether they are nucleated, linear or dispersed

Shanty/informal settlement/favela/ bustee Terms used in different parts of the world for a slum dwelling, which is a temporary or poorly built construction, often on land that does not belong to the householder

Sheet wash Removal of a thin surface layer of soil in a sheet of rainwater flowing over a gentle slope

Shingle Stones, more than 2 mm in diameter, rounded by attrition, found on beaches

Short-haul Travel to a destination in less than three hours

Site The land that a settlement is built on

Situation The position of a settlement in the surrounding area

Slum Houses considered not fully fit for habitation

Small-scale map A map which shows a relatively large area of land in less detail, for example a map of the world

Smog Fog mixed with pollutants, such as smoke and sulfur dioxide

Social factor An influence of the beliefs and practices of a community of people

Soil erosion The removal of soil by rain or wind

Soil exhaustion Loss of soil fertility by being depleted of its nutrients

Soil fertility How enriched the soil is with nutrients

Soil structure The soil's individual particles are held together by decayed organic matter (humus)

Solid biofuels Often referred to as biomass, it can be used in power stations and in the heating systems of houses and buildings

Solution (erosion) The dissolving of calcium carbonate and other soluble minerals in sea/river water

Solution (transport) Transport of material dissolved in the water

Spit Long, narrow, low ridge of sand or shingle deposited at a bend in the coast, such as a bay or river mouth. Spits are attached to the land at one end and the other ends in open water. They may have re-curved or hooked ends

Spot heights A dot on the map with a number beside it to show the number of metres that the point indicated is above sea level

Spurs A ridge which slopes down from high ground to low ground. A spur is shown by a V-shape in the contours, where the V points to low ground

Standard of living To do with money and wealth

Stomata A pore in the leaf surface layer through which water is transpired

Storm surge A rapid rise of sea level (caused by lower air pressure), which is driven onshore by strong winds as a high, powerful mass of water

Straight Extending or moving in one direction, without a curve or bend

Stratified sample A sample that divides the population into subgroups and represents each one in the same proportions that it has in the population. The same principle is followed where the differences are in area

Stratocumulus Low-level cloud made of water droplets. Stratus with heaped sections, giving white and grey parts

Stratus Low-level cloud made of water droplets. A thin, uniform, grey sheet with a fairly flat base. It may be thick enough to produce drizzle

Subduction The process where plates converge and one plate is forced beneath the other

Subsistence farming The farmer grows crops or rears animals for consumption for his or her family. A surplus may be produced from time-to-time, which is sold

Surface water Water from rivers and lakes

Suspension Process of transport in which mud and silt is held within the body of the water, making it discoloured

Sustainable development An activity done in such a way that the resources it uses will be available to future generations and the activity will be able to continue

Systematic sample Data collected in a regular order, e.g. every tenth house or person

Tally chart Recording method for a count by which a mark is made in a line for each observation, with each fifth line crossing the first four at a diagonal angle

Temperature The degree of heat or coldness of the air

Temperature range The calculated difference between the highest and lowest temperatures for the period

Tertiary industry Industry that provides a service, for example banking

Thermal power Electricity produced by burning fossil fuels

Threshold population The minimum number of people needed to provide a large enough demand for a service

Throughflow Water flowing through the soil

Tourist A person who travels to another country for more than a day but less than a year

Traction Process of transport in which material rolls downstream staying in contact with the riverbed

Transect A line along which samples are taken

Transnational corporation (TNC) A large company that operates (as a producer or seller) in many countries or continents. These companies control an increasing proportion of the global economy

Tributary A smaller river joining a larger one

Trigonometrical point This is a pillar about a metre tall which is used as a fixed point by the mapmakers. It is shown on a survey map along with its height above sea level

Tropical storm An intense area of low pressure that forms in the tropics and moves pole-wards, bringing violent winds and intense rainfall to coastal areas in the tropics and subtropics

Tsunami An ocean wave produced when there is movement of the seabed by the fault movement, which causes an earthquake. Also, caused by the collapse of a volcanic cone into the sea

Under-population There are too few people to use all the resources of a country to maximum efficiency

Urban Associated with towns and cities

Urbanization The increase in the percentage of the population living in towns and cities

Vent The pipe which feeds magma to the surface

Water deficit Where water demand exceeds supply

Water surplus Where supply exceeds demand

Watershed The dividing line between two drainage basins

Well A vertical shaft, a metre or so wide, sunk into rock to extract groundwater, sometimes using buckets

Wet point A site with reliable sources of water from rivers, springs and wells in an otherwise dry area

Revision planners

	Revision Period 1	Revision Period 2	Revision Period 3	Revision Period 4	Revision Period 5	Revision Period 6
Sunday						
Saturday						
Friday						
Thursday						
Wednesday						
Tuesday						
Monday						

	Revision Period 1	Revision Period 2	Revision Period 3	Revision Period 4	Revision Period 5	Revision Period 6
Monday						
Tuesday						
Wednesday						
Thursday						
Friday						
Saturday						
Sunday						